EVERYDAY MAGIC IN EARLY MODERN EUROPE

Everyday Magic in Early Modern Europe

Edited by

KATHRYN A. EDWARDS
University of South Carolina, USA

Routledge
Taylor & Francis Group

LONDON AND NEW YORK

First published in paperback 2024

First published 2015 by Ashgate Publishing

Published 2016 by Routledge
4 Park Square, Milton Park, Abingdon, Oxon OX14 4RN

and by Routledge
605 Third Avenue, New York, NY 10158

Routledge is an imprint of the Taylor & Francis Group, an informa business

Publisher's Note
The publisher has gone to great lengths to ensure the quality of this reprint but points out
that some imperfections in the original copies may be apparent.

British Library Cataloguing in Publication Data
A catalogue record for this book is available from the British Library

The Library of Congress has cataloged the printed edition as follows:
Everyday magic in early modern Europe / edited by Kathryn A. Edwards.
    pages cm
  Includes index.
  ISBN 978-1-4724-3350-3 (hardcover : alk. paper)
  1. Magic—Europe—History. 2. Occultism—Europe—History.
  3. Europe—Social life and customs. 4. Europe—Folklore. I. Edwards, Kathryn A., 1964– , editor.
  BF1595.E94 2015
  133.4'3094–dc23

                                                                        2015015240

ISBN: 978-1-4724-3350-3 (hbk)
ISBN: 978-1-03-292807-4 (pbk)
ISBN: 978-1-315-58133-0 (ebk)

DOI: 10.4324/9781315581330

Contents

Notes on Contributors

Jason Coy is Associate Professor at the College of Charleston (USA) and received his doctorate at the University of California, Los Angeles, in 2001. He is the author of *Strangers and Misfits: Banishment, Social Control, and Authority in Early Modern Germany* (Brill, 2008). He is also co-editor, with Benjamin Marschke and David Sabean, of *The Holy Roman Empire, Reconsidered* (Berghahn Press, 2010) and, with Jared Poley, Benjamin Marschke, and Claudia Verhoeven, of *Kinship, Community, and Self: Essays in Honor of David Warren Sabean* (Berghahn Press, 2014). He is writing a book that explores divination in Germany between 1500 and 1800.

Johannes Dillinger is Professor of Early Modern History at Oxford Brookes University (England) and honorary Professor of Modern History and Regional History at Johannes Gutenberg University Mainz (Germany). He specializes in early modern European magic, constitutional history, and comparative history. Among his most recent books are *Kinder im Hexenprozess. Magie und Kindheit in der Frühen Neuzeit* (Steiner, 2013); *Magical Treasure Hunting in Europe and America: A History* (Palgrave, 2012) [*Auf Schatzsuche* (Herder, 2011)]; and *"Evil People": A Comparative Study of Witch Hunts in Swabian Austria and the Electorate of Trier* (University of Virginia, 2009). He is currently editing the *Routledge History of Witchcraft*.

Kathryn A. Edwards is Professor of History at the University of South Carolina (USA). She has published several books including *Leonarde's Ghost: Popular Piety and "The Appearance of a Spirit" in 1628* (Truman State University Press, 2008; coauthored with Susie Speakman Sutch), *Werewolves, Witches, and Wandering Spirits: Folklore and Traditional Belief in Early Modern Europe* (Truman State University Press, 2002; editor), and *Families and Frontiers: Family and Communal Re-creation in the Early Modern Burgundies* (Brill, 2002). Her current research is a synthetic history of European beliefs about ghosts from the late Middle Ages to the early twentieth century; the first volume of that research is titled *Living with Ghosts: The Dead in European Society from the Black Death to the Enlightenment*.

Sarah Ferber is Associate Professor of History at the University of Wollongong (Australia). She is the author of *Demonic Possession and Exorcism in Early*

Modern France (Routledge, 2004). She contributed several entries to Richard M. Golden, ed., *Encyclopedia of Witchcraft* (ABC-CLIO, 2004) and wrote the chapter on demonic possession for Brian P. Levack, ed., *The Oxford Handbook of Witchcraft in Early Modern Europe and Colonial America* (Oxford University Press, 2013). She is currently preparing a monograph providing an overview of magic, witchcraft, and demonology in pre-modern Europe and its colonies.

Linda Lierheimer is Professor of History and Humanities at Hawaii Pacific University (USA). She edited and translated *The Life of Antoinette Micolon* (Marquette University Press, 2004), and her article, "Gender, Resistance, and the Limits of Episcopal Authority: Sébastien Zamet's Relationships with Nuns (1615–1655)," appears in Jennifer M. DeSilva, ed., *A Living Example: Episcopal Reform, Relations, and Politics in Early Modern Europe* (Truman State University Press, 2012). She is currently working on a book about conflicts between nuns and bishops in seventeenth-century France.

Doris Moreno Martínez is Professor of Early Modern History at the Universitat Autònoma de Barcelona (Spain). She has worked on the Inquisition, elite culture, and history of the Jesuits in the early modern period, and has organized conferences and published extensively on these topics. She has edited, with Alexandre Coello de la Rosa, *Jesuitas e imperios de ultramar, siglos XVI–XX* (Sílex, 2012) and has worked in collaboration with Ricardo García Cárcel on *Inquisicion. Historia crítica* (Temas de hoy, 2000). She is the author of *La invención de la Inquisición*, (Marcial Pons, 2004) and has published *Protestantes, visionarios, profetas y místicos* (Debolsillo, 2004) with A. Fernández Luzón. She is currently preparing a book on the Spanish origins of the Black Legend about the Society of Jesus.

Antoine Mazurek is a postdoctoral researcher at the Institut d'Histoire Moderne et Contemporaine (Université de Paris, Panthéon-Sorbonne—Ecole Normale Supérieure—CNRS). He received his PhD in History at the École des Hautes Etudes en Sciences Sociales (France) where his doctoral thesis (2013) on "The Guardian Angel in the Modern Era: Worship, Customs, and Doctrinal Development, Sixteenth to Eighteenth Centuries" ("L'ange gardien à l'époque moderne: culte, élaboration doctrinale et usages, XVIe–XVIIIe siècles") earned a "with distinction." It will appear with Les Belles Lettres. He has also published the article "Angel" in the *Dictionnaire des faits religieux* (Presses universitaires de France, 2010) and "Réforme tridentine et culte des saints en Espagne: liturgie romaine et saints ibériques," in *The Council of Trent: Reform and Controversy in Europe and Beyond (1545–1700)*, actes du colloque des 4-6 déecembre 2013 tenu à Louvain (Vandenhoeck & Ruprecht, forthcoming).

Jared Poley is Associate Professor of History at Georgia State University (USA). He is the author of *Decolonization in Germany: Weimar Narratives of Colonial Loss and Foreign Occupation* (Peter Lang, 2005) and *A Modern History of Greed: Religion, Economics, Health* (forthcoming). Poley is also co-editor, with David M. Luebke, Daniel C. Ryan, and David Warren Sabean, of *Conversion and the Politics of Religion in Early Modern Germany* (Berghahn Books, 2012) and, with Jason Coy, Benjamin Marschke, and Claudia Verhoeven, of *Kinship, Community, and Self: Essays in Honor of David Warren Sabean* (Berghahn Books, 2014).

Raisa Maria Toivo is Research Fellow at the Finnish Academy Center of Excellence "Re-thinking Finland: History of a Society 1400–2000" at the University of Tampere (Finland). Her publications include *Witchcraft and Gender in Early Modern Society: Finland and the Wider European Experience* (Ashgate, 2008). She has also edited, with Marko Nenonen, *Writing Witch-Hunt Histories* (Brill, 2014) and, with Marianna Muravyeva, *Gender in Late Medieval and Early Modern Europe* (Routledge, 2012). She is currently working on religious pluralism in early modern Finland and Europe.

Chapter 1

Introduction:
What Makes Magic Everyday Magic?

Kathryn A. Edwards

The last decades of the twentieth and the beginning of the twenty-first centuries have seen an explosion of research on magical practices and the attitudes about them in late medieval and early modern Europe. Inspired by books as diverse as Keith Thomas's *Religion and the Decline of Magic* and Francis Yates's *The Occult in the Elizabethan Age*,[1] scholars have combined methods from anthropology, literary studies, criminology, psychology, and gender studies to arrive at a more profound understanding of what has been described as the "enchanted world-view" of this time.[2] In the process work on witchcraft has dominated. The early modern witch hunts provide a dramatic and, to modern European sensibilities, abhorrent example of one consequence of belief in a magical and immanent world. Strangely compelling, the causes, ideologies, and practices of Europe's witch hunts have been frequently debated, often revealing a great deal

---

[1]   Keith Thomas, *Religion and the Decline of Magic: Studies in Popular Belief in Sixteenth- and Seventeenth-Century England* (London: Weidenfeld and Nicolson, 1971); Francis Yates, *Occult Philosophy in the Elizabethan Age* (London: Routledge & Kegan Paul, 1979).

[2]   Among the individual monographs that have most influenced my perspective on Europe's witch hunts see Bengt Ankarloo and Stuart Clark, eds, *The Athlone History of Witchcraft and Magic in Europe*, 6 vols (London: 1996–2002); Wolfgang Behringer, *Witchcraft Persecutions in Bavaria: Popular Magic, Religious Zealotry and Reason of State in Early Modern Europe* (Cambridge: Cambridge University Press, 1997); Robin Briggs, *The Witches of Lorraine* (Oxford: Oxford University Press, 2007); Stuart Clark, *Thinking with Demons: The Idea of Witchcraft in Early Modern Europe* (Oxford: Oxford University Press, 1997); Lyndal Roper, *Oedipus and the Devil: Witchcraft, Religion, and Sexuality in Early Modern Europe* (London: Routledge, 1994); Jonathan Seitz, *Witchcraft and Inquisition in Early Modern Venice* (Cambridge: Cambridge University Press, 2011); James Sharpe, *Instruments of Darkness: Witchcraft in England, 1550–1750* (Harmondsworth: Penguin, 1997). The most comprehensive recent syntheses are Brian P. Levack, ed., *The Oxford Handbook of Witchcraft in Early Modern Europe and Colonial America* (Oxford: Oxford University Press, 2013); David J. Collins, S.J., ed., *The Cambridge History of Magic and Witchcraft in the West: From Antiquity to the Present* (Cambridge: Cambridge University Press, 2015).

about their authors' sympathies or frustration with the past. The availability of records have helped this academic focus; the records of witchcraft trials often appear in relatively coherent collections, more accessible for analysis, than many other records of magical practices. Information about witches' activities, characteristics, and relationships can be found throughout demonologies, trial records, and lists of sentences. In the process of studying such texts, scholars have traced personal, professional, and intellectual networks at all levels of European society. They have also described a world that magic pervaded.[3]

In so doing, scholars have struggled with an admittedly imprecise vocabulary whose meanings could vary in late medieval and early modern Europe and have certainly shifted between that time and the present. "Superstition," "the supernatural," and "the natural" were all concepts key to understanding witchcraft and magic more generally but on which scholars have taken years to come to common definitions—definitions that are still ignored, unfortunately, in many less specialized works.[4] What to call common beliefs and practices has also been widely disputed, and most terms have been faulted in some way for presumed anachronism. "Traditional religion," "common religion," "popular religion," "shared religion," and "lived religion" have all been critiqued in some form, although since no acceptable substitute has been agreed to, most scholars use one of the above terms or some minor variant to signal beliefs and practices that were widely shared among classes, genders, and educational levels.[5]

---

[3]   Recently many historians studying witchcraft and magic have used Max Weber's concept of "disenchantment" as applied to Europe's social and intellectual changes in the seventeenth and eighteenth centuries as a touchstone for these debates. For key contributions to this debate, see Alexandra Walsham, "The Reformation and 'The Disenchantment of the World' Reassessed," *The Historical Journal* 51:2 (June 2008): 497–528; Euan Cameron, *Enchanted Europe: Superstition, Reason, and Religion in Europe, 1250–1750* (Oxford: Oxford University Press, 2010); Andrew Keitt, "Religious Enthusiasm, the Spanish Inquisition, and the Disenchantment of the World," *Journal of the History of Ideas* 65:2 (April 2004): 231–50; H.C. Erik Midelfort, *Exorcism and Enlightenment: Johann Joseph Gassner and the Demons of Eighteenth-Century Germany* (New Haven: Yale University Press, 2005).

[4]   Michael D. Bailey has made several convincing and comprehensive analyses of aspects of the debate over "superstition": *Fearful Spirits, Reasoned Follies: The Boundaries of Superstition in Late Medieval Europe* (Ithaca: Cornell University Press, 2013); "A Late-Medieval Crisis of Superstition?" *Speculum* 84:3 (July 2009): 633–61; "The Disenchantment of Magic: Spells, Charms, and Superstition in Early European Witchcraft Literature," *The American Historical Review* 111:2 (April 2006): 383–404. Also see Helen Parish and William Naphy, eds, *Religion and Superstition in Reformation Europe* (London: Palgrave Macmillan, 2002), introduction, and David J. Collins, "Introduction," in Collins, *The Cambridge History of Magic and Witchcraft in the West*, 1–14.

[5]   For the problems in finding satisfactory terms, see Kathryn A. Edwards, "Popular Religion," in   *Reformation and Early Modern Europe: A Guide to Research*, ed. David M. Whitford (Kirksville, MO: Truman State University Press, 2007) 331–54; Eamon Duffy,

In some cases these attitudes and activities transcended confessions, while in others they were confined to a village or region. They followed a spectrum of what theologians saw as orthodox, ranging from saying common prayers (Ave Maria, Pater Noster) to, among Catholics, sprinkling salt to purify an area. This spectrum also slid into the suspect, such as using the Bible as a protective amulet, and could even become heresy or witchcraft, as it did when a person called on the Virgin or saints for aid in transferring milk from a neighbor's cow to their own. Despite the very real differences in such practices, however, they reflected a continuum of beliefs where supernatural influence and access were presumed. They could be remarkable or ordinary, even mundane, but they attested to a shared and durable sense of how the world worked, humanity's place in it, and the relationship between forces not of this world, extraordinary forces in this world, and daily experiences.[6] This is the "everyday" of this book's title and chapters.

The definitions of and distinctions between "magic" and "religion" have also been at the heart of many of these debates, and the wobbly ground between the two is where this book is situated. Older historiography relied on a distinction drawn from anthropology between magic as operational, its practitioners focused on control and results, and religion as supplicatory, where believers asked God for assistance rather than demanding it.[7] The same historiography that has contributed so much to our understanding of witchcraft and magical practices more generally has shown that such a distinction was untenable for late medieval and early modern Europe. While learned necromantic practices fall clearly into both modern and pre-modern concepts of magic, what made other actions magical frequently depended on who assessed them.[8] If a priest, followed by villagers, carried a monstrance containing a Eucharistic wafer three times around a field to ensure a good harvest, was that magic? The priest and the villagers alike

---

The Stripping of the Altars: Traditional Religion in England, 1400–1580 (New Haven: Yale University Press, 2005), introduction; Jacques Berlinerblau, "Max Weber's Useful Ambiguities and the Problem of Defining 'Popular Religion,'" *Journal of the American Academy of Religion* 69:3 (Sept. 2001): 605–26.

6   Benedicta Ward has made a similar argument for miracles in medieval Europe: *Miracles and the Medieval Mind: Theory, Record, and Event, 1000–1215* (Philadelphia: University of Pennsylvania Press), 216.

7   See Thomas, *Religion and the Decline of Magic*; Ronald Hutton, "Anthropological and Historical Approaches to Witchcraft: Potential for a New Collaboration?" *The Historical Journal* 47, no. 2 (2004): 413–34.

8   Although he might disagree with the scope of the definition of magic provided here, Richard Kieckhefer has demonstrated the impressive range of medieval magical practices and inspired, in part, this interpretation: *Magic in the Middle Ages* (Cambridge: Cambridge University Press, 1989); *Forbidden Rites: A Necromancer's Manual of the Fifteenth Century* (Stroud: Sutton, 1997); "Magic and its Hazards in the Late Medieval West," in Levack, *The Oxford Handbook*, 13–31.

would almost certainly say that it was religion, while modern observers might argue that it was magic. The scholars in this collection would likely argue that it was both, that the boundaries between religion and magic, licit and illicit belief and practice were porous and fluid.[9] If modern scholars classify that practice as magic (and often, therefore, seen as illegitimate) or religion (and, therefore, legitimate) is beside the point. Such processions were founded on the belief in a pervasive enchantedness and wonder that most Europeans accepted at this time. Whether this enchantedness accessed the supernatural, the preternatural, or the wondrous, it relied on a continuum of beliefs and practices where marvelous and miraculous experiences were uncommon, but they were expected and accepted, and life was generally interpreted within this framework.[10] This worldview and its attendant activities is the "magic" of this collection's title.

Given these perspectives, "everyday magic" can encompass an enormous number of topics in late medieval and early modern European history, as the recent work of Stephen Wilson on Europe's "magical universe" has shown.[11] It can also bridge modern neuropsychology and early modern villages.[12] This collection is by no means as comprehensive. Instead, it uses studies of particular magical activities, theories about magic, and encounters with the magical and marvelous to contribute to and reevaluate broader currents in European historiography in which the pervasiveness of a "magical" perspective may

---

[9]   A similar argument has been made for the definition of witchcraft by Clark and Ankarloo, "Introduction," vol. 3 *Athlone History of Witchcraft*, x.

[10]   The supernatural is generally understood as that which is from outside this world (generally divine) and beyond human understanding, the preternatural as that which is of the world but beyond human understanding, and the wondrous as that which is of the world and capable of being understood by humans but is beyond normal experience and comprehension. For a more detailed discussion of magic's meanings, see Michael D. Bailey, "The Meanings of Magic," *Magic, Ritual, and Witchcraft* 1:1 (Summer 2006): 1–23; Owen Davies, *Magic: A Very Short Introduction* (Oxford: Oxford University Press, 2012), 1–13.

[11]   Stephen Wilson, *The Magical Universe: Everyday Ritual and Magic in Pre-modern Europe* (London: Hambledon, 2000). Wilson integrates materials from the thirteenth to the nineteenth centuries, often in the same or neighboring paragraphs, arguing that they stem from a remarkably constant, pre-modern world-view.

[12]   Edward Bever, *The Realities of Witchcraft and Popular Magic in Early Modern Europe: Culture, Cognition, and Everyday Life* (London: Palgrave Macmillan, 2008) focuses more on the psychological and neurophysiological foundations for such popular beliefs, as does his "Current Trends in the Application of Cognitive Science to Magic," *Magic, Ritual, and Witchcraft* 7:1 (Summer 2012): 3–18. His approach has been somewhat controversial: see the forum, "Contending Realities: Reactions to Edward Bever," in *Magic, Ritual, and Witchcraft* 5:1 (Summer 2010): 81–121 (contributors include Michael D. Bailey, Stuart Clark, Richard Jenkins, Rita Voltmer, Willem de Blécourt, Jesper Sørensen, and Edward Bever); Malcolm Gaskill, "The Pursuit of Reality: Recent Research into the History of Witchcraft," *The Historical Journal* 51, no. 4 (2008): 1069–88.

not always be appreciated. It integrates material that is generally placed in separate books on magic and on piety so that its authors can argue for a more inclusive definition of "everyday magic." Beginning with chapters that compare programmatic statements to lived experience, it then alternates chapters on the theoretical underpinnings of or academic debates over supernatural entities and events that were perceived as affecting daily life with chapters on the lived experiences that developed from such events and interactions with such entities. Both perspectives are key to understanding everyday magic. For example, as the chapters here will show, the activities theorized for guardian angels bore an uncanny resemblance to those of the spirits helping treasure hunters, and spirits and forces of all kinds—not merely demonic or angelic—could be called on equally for help in divination and prognostication, activities that fell well within a traditional definition of magic. To paraphrase the final chapter, the demonic could even be domesticated, that is, it could become an aspect of everyday magic. As these chapters demonstrate, such domestication was contingent on many local factors and at times failed. In all cases, though, it evolved with the societies and belief systems in which it was embedded. The studies here present aspects of that evolution in everyday magic and place such magic within a spectrum of pious practices and daily activities.

* * *

As noted above, the chapters collected here consider several of the broad historiographic questions at the heart of studies of witchcraft, magic, and popular belief. They also suggest that other themes in early modern European history may deserve a greater role in this historiography than they presently enjoy. In the process, certain topics recur throughout many chapters. Rather than summarizing each chapter individually, then, I will focus here on the themes binding these articles and their contribution to discussions of everyday magic.

Each author is well aware of the debates over the definition of magic and superstition and the challenges in distinguishing between modern and pre-modern perspectives. The chapters by Antoine Mazurek, Doris Moreno Martínez, Jason Coy, and Jared Poley show that early modern theologians and jurists from both Protestant and Catholic confessions continued to share the late medieval definition of superstition as excessive and improper religious observance. As these authors note, such "superstitious" practices could cause ecclesiastical hierarchies as much, if not more, concern than witchcraft well into the sixteenth and seventeenth centuries. That clergy and laity could share such superstitions made them all the more fraught; the commonality of certain magical practices show that people of all ages, classes, genders, and educational levels could accept and advocate for superstition. The very qualities that made something magical could vary from circumstance to circumstance, as Raisa

Maria Toivo shows for early modern Finland, a situation that made defining magic and eradicating its superstitious elements almost impossible.[13] There, farmers could interpret the same action designed to protect their cattle as either skill or magic, depending on who was doing it. If the action was deemed magical, it *de facto* became superstitious and potentially dangerous, yet if it was proven to be a skill, then the farmer would be remiss not to use it. In fact, abandoning that skill could even be seen as slighting God's gifts by ignoring a valuable tool he provided for the farmer's benefit.

As the case of the Finnish farmers demonstrates, magic itself could have many aspects and implications; the same individual who caused illness could often provide the means for its cure as well as create protective amulets, find missing objects, or prepare love potions. Owen Davies has rightly noted that the line between a practitioner of everyday, common magic (a cunning person) and a witch could be fine indeed, but seemingly it was well understood in early modern Europe.[14] Moreover, cunning people could enjoy lucrative and long careers at the same time that others were being accused of and prosecuted for witchcraft. The chapters by Johannes Dillinger and Raisa Maria Toivo contribute to this discussion about everyday magical practitioners. Through analysis of trial records from early modern Germany and Finland, they highlight the cultural contingency in distinguishing between the qualities of magic: good, bad, or somewhere in the middle. In his study of treasure hunting, Dillinger argues that a key component in making magic dangerous and suspect was the harm that it might cause or be perceived to cause others, a version of the *maleficia* so familiar to scholars of witchcraft. Toivo, however, stresses that, for early modern Finns, classifying magic was unimportant, and the words associated with magic changed their connotation over time and depending on confessional circumstances.

Such magical powers depended on an ability to manipulate natural or supernatural forces, and Mazurek's article focuses on how these forces were interpreted in early modern Spanish and Italian debates over the role of the guardian angel. Special knowledge and substance often made angels and demons central figures in magical practices ranging from elite necromancy to more common divination and protective spells. By their own nature, both angels and demons understood and could manipulate the created world far better than humanity, and both were willing to do so, according to the theologians Mazurek analyzes. For these intellectuals, guardian angels thus provided an invaluable bridge between the natural environment and troubles with which humans had

---

[13]   For an assessment of how such ambiguities affect witchcraft historiography, see her "The Witchcraze as Holocaust: The Rise of Persecuting Societies," in *Witchcraft Historiography*, eds. Jonathan Barry and Owen Davies (Houndsmills: Palgave, 2007), 90–107.

[14]   Owen Davies, *Popular Magic: Cunning-Folk in English History* (London: Bloomsbury Academic, 2007).

to cope and the salvation humans hoped to obtain. Yet the activities guardian angels inspired humans to perform could appear as magical and, by implication, suspect, either through the way angels could manipulate natural forces or the support angels gave to human attempts to protect themselves from demonic temptation. The visionary nuns of Linda Lierheimer's chapter on false sanctity were subject to similar analyses. Their spiritual guides assessed if their impulses were demonically or divinely (angelically) inspired in part by the balance of natural, unnatural, and supernatural behaviors the nuns exhibited. Claims of angelic assistance could actually undermine a visionary's status if the angel inspired her to behave in a way that did not exhibit the qualities of spiritual perfection that were expected in a woman under angelic guardianship.

Lierheimer's chapter is one of many in this collection that insist on the importance of local, historicized analysis to understand the attitudes towards and practices of everyday magic in early modern Europe. Such a focus should not surprise anyone familiar with the historiography on witchhunting. There, local circumstances—law codes, judicial structures, public health, social networks, and neighborhood relationships—have been central to analyses for decades. In subjects like everyday magic, however, the degree of historicization can vary widely. For decades, even centuries, folklorists have gathered and organized material on everyday magic and beliefs. Often they have classified these materials thematically and analyzed them to show broad cognitive and cultural patterns linked to a relatively constant pre-modern, folk, or subconscious sensibility. While such work can be invaluable, historians must use it cautiously if they want to assess changes over time, the influence of the cultural environment, and the relationship between such beliefs and other historical trends.[15] The chapters by Toivo, Moreno, Lierheimer, Dillinger, and Edwards all demonstrate this cautious integration of folkloric materials into historicized analysis. They embed everyday magical practices and experiences within a historically specific social and cultural context. In the process a broad cast of characters is seen to be involved in magic—men and women, nuns and housekeepers, theologians and farmers—and the types of magic that are accepted as "everyday magic" and individuals' responses to its use varies. Farmers see magic as a necessary tool, treasure hunters present their hunt in a sacramental framework, and Huguenot ministers calmly question ghosts. Although magic involved cooperation in some cases, as Lierheimer and Dillinger show, contests for the right to exhibit magical and mystical effects could also divide communities. Rather than existing in a consistent and enduring folkloric framework, the constitutive characteristics of everyday magic evolved to reflect local circumstances and, as Sarah Ferber argues, in dialog with religious and secular authorities.

---

[15] Among the better recent work that reflects this more folkloric and thematic interpretive style, see Wilson, *The Magical Universe* and Davies, *Popular Magic.*

By emphasizing local circumstances, such work demands an appreciation of daily, personal concerns for early modern Europeans. Rather than focus on power and ideology, then, the articles in this collection especially stress the economic role and benefits of everyday magic. Toivo and Moreno develop the more familiar ground of the damage magic could do to an individual's livelihood, stressing the pragmatism and empiricism of magical practitioners, while Dillinger and Poley situate money—actual cash in gold and silver—within a magical world. Neither stresses the financial benefits due to a magical practitioner, be it a cunning person, wandering exorcist, or witch. Rather Poley studies the interpretation of dreams: money as a symbol of temptation and the ways its prominence reflected everyday concerns during the time of the Price Revolution. For Dillinger, the pursuit of treasure (money and small, portable valuables) links common magical practices with local and even regional trade in the figure of the treasure hunter. Stressing the expense and distance involved in finding some books that treasure hunters regarded as essential to their task, Dillinger portrays figures that are simultaneously pragmatic and idealistic. Like Poley, Dillinger also situates his discussion of magic in broader economic concepts—in his case, that of "the limited good"—and he uses that concept to describe why some magical practices were regarded as licit and others as illicit.[16]

Not surprisingly, an underlying theme of many articles in this book is the effect of the religious divisions in the sixteenth and seventeenth centuries. More specifically, how were confessional differences reflected in magical practices and the response of communities and authorities to them? How did Protestant communities especially—here Lutheran and Calvinist, although interesting work has been done on Dissenters and witchcraft[17]—deal with the tension between theological directives and traditional practice? Did their approach to magic change as the Reformation became more established? Mazurek and Coy, in particular, assess clerical attempts to direct what they perceive as common and dangerous beliefs, showing Catholic and Calvinist responses, respectively. Timing becomes especially important for Bartholomaeus Anhorn, the protagonist of Coy's chapter. Working in the late seventeenth century, Anhorn faced a growing group of influential individuals who were skeptical about the prevalence and effectiveness of magic and the supernatural in daily life. Anhorn's defense of demonological truisms and condemnations of magical practices illustrate the problems Reformed theologians faced when

---

[16]   The influence of fear and the limited good on witch trials and magical practice more generally underlies Malcolm Gaskill's brief survey: *Witchcraft: A Very Short Introduction* (New York: Oxford University Press, 2010), esp. 39–45.

[17]   Gary K. Waite, *Eradicating the Devil's Minions: Anabaptists and Witches in Reformation Europe* (Toronto: University of Toronto Press, 2007); Garry K. Waite, "Sixteenth-Century Religious Reform and the Witch-Hunts," in Levack, *The Oxford Handbook*, 485–506.

they became the religious "establishment." Toivo and Dillinger consider the situation facing Protestants who continued to use magic based on traditional, often Catholic, precepts and practices as a means of assessing the influence of confessionalization. Expanding that theme, Edwards examines the reactions of a Huguenot minster who is cursed and whose house is haunted by a spirit that even he has trouble classifying.

The prevalence and diversity of spirits in lived experience and everyday magic is a recurring theme and one that several articles argue should be integrated more thoroughly into broader studies of popular belief. Certainly the idea that demons and angels populated this world is common in histories of early modern Europe, but spirits could be far more diverse than that binary suggests. Alongside such theologically acceptable entities were fairies, dwarves, will-o'-the-wisps, flying serpents, white ladies, spectral hunters, and ghosts themselves.[18] Each of these figures played a role in the "magical universe" of early modern Europe, and each could influence everyday magic. In their discussion of spiritual discernment, both Mazurek and Lierheimer describe a subject in which spirits are often conceived as starkly demonic or angelic; they show that some of the most revealing situations could arise when those involved feared that some other type of spirit could inspire a vision. For Edwards, the ambiguity of a spirit's status is the heart of the matter, and the attempts (magical and otherwise) to deal with such a visitation reveal the imperfect incorporation of Calvinist orthodoxy even by a clerical authority. Diverse spirits, especially ghosts, are central to the magical practices involved in treasure hunting, the subject of Dillinger's article. Not only did ghosts inspire some hunters to find treasure and return it—they saw it as a work of mercy—but magical beings and mysterious spirits could guard the treasure hunters, requiring treasure hunters to use both magical and pious means to expel them.

As these examples suggest and the more detailed studies to follow will argue, magical events and attitudes towards them need to be seen as part of a continuum; given the perceived pervasiveness of the preternatural and supernatural in early modern Europe, both the bad and the beneficent were regarded through similar theoretical and social lenses. In the process, the authors question the extent to which magic, and the supernatural more generally, were believed to be dangerous. While both certainly could be fraught, the sources of information on which modern scholars rely, such as trial records and condemnatory treatises, often

---

[18] Recent work has explored some of these permutations: Timothy Chesters, *Ghost Stories in Late Renaissance France: Walking by Night* (Oxford: Oxford University Press, 2010); Miriam Rieger, *Der Teufel im Pfarrhaus: Gespenster, Geisterglaube und Besessenheit im Luthertum der Frühen Neuzeit* (Stuttgart: Franz Steiner, 2011); Lizanne Henderson, *Scottish Fairy Belief: A History* (Edinburgh: John Donald, 2007); Diane Purkiss, *At the Bottom of the Garden: A Dark History of Fairies, Hobgoblins, Nymphs, and Other Troublesome Things* (New York: New York University Press, 2003).

make magic seem more threatening than it was thought to be. Appreciating the everyday, even mundane, aspects of magic and a magical worldview can thus lead to a deeper understanding of the debates over licit and illicit magical practices and of early modern European society more generally.

Chapter 2

Magical Lives: Daily Practices and Intellectual Discourses in Enchanted Catalonia during the Early Modern Era

Doris Moreno Martínez[1]

Challenging the traditional distinction between learned and popular culture or the division between learned or official religiosity and popular religiosity, historian and anthropologist Julio Caro Baroja underlined "the complexity of religious life" in early modern Spain. Instead of offering simplified dichotomies, Caro Baroja argued for the need to study forms and practices instead of discourses and norms and to appreciate the world in which these practices were taking place (thus, his definition of "the religious life"), specifying that the people of the Old Regime only experienced life in a religious way. Understood generally as the presence of the supernatural and the transcendental, religion had a deep reality in early modern Spain, as Caro Baroja emphasized. The magical, the supernatural, and everyday life were one.[2] Similar circumstances existed in early modern Catalonia, where the magical, the supernatural, and everyday life were also one, although not the same.

In the process of analyzing these circumstances, this article nuances two common conceptions about pre-modern religious life and its integration into everyday practices. The first appears in the work of historians like Jean Delumeau and Michel Vovelle who have distinguished two elements in a

---

[1]   ORCID: 0000—0003—2880—9533. The following research projects were involved in supporting this article; the Ministry of Economy and Finance RyC-2008–02585 and HAR2011–26002; and the Government of Catalonia 2014SGR1206. I would especially like to thank the translators, Maria Soukkio and Kathryn A. Edwards.

[2]   Julio Caro Baroja, *Las formas complejas de la vida religiosa* (Madrid: Akal, 1978) and *Vidas mágicas e Inquisición*, 2 vols. (Madrid: Istmo, 1992; 1st ed., 1967). A historiographical reflection on this author and his influential work on this topic was the subject of a volume of *Historia Social* 55 (2006), especially the articles by Manuel Peña Díaz, "Caro Baroja y la religiosidad en la España del Siglo de Oro," 25–44 and José Luis Betrán, "El mundo mágico de Julio Caro Baroja," 79–112. Fundamental for a global perspective is Stephen Wilson, *The Magical Universe: Everyday Ritual and Magic in Pre-Modern Europe* (London: Hambledon, 2004).

religious worldview deeply influenced by all things supernatural: a pagan substrate and/or Christianity. The scarcely Christianized Europe of the Middle Ages was followed by an early modern period during which Catholics and Protestants fought over territory and domination of a religious market that included catechizing and reform of the rural population who had been left out of the Christianization process. According to scholars who support these interpretations, a symbolic fracture zone between pre-Christian conceptions and this early modern belligerent Christianity occurred when witch hunts broke loose in central Europe during the seventeenth century. By the eighteenth century, Europe had been increasingly de-Christianized, which, in turn, resulted in the birth of revolutionary ideals. Today the aforementioned claims are under revision; modern scholars argue that early modern Christianization was not as successful as previously thought, nor was the de-Christianization of the eighteenth century as intense as previously claimed.[3]

The second conception approaches this topic from the point of view of studies on power. In those works the history of religion in early modern Europe has been connected with the development of the modern, confessional state that used the church to impose political and social order and to establish a disciplined society submissive to that established power. In the Catholic world, the Council of Trent and the Inquisition were excellent companions for the political powers, the former applying its policies and persuasive strategies to achieve the long desired Catholic Reform and the latter serving as the "armed wing" of the Catholic Church.[4] To this "top-down" vision, it is necessary to juxtapose a bottom-up perspective, as have historians critical of the concepts of confessionalization and social discipline. Such critics argue that there were many

---

[3]　Jean Delumeau, *El catolicismo de Lutero a Voltaire*, trans. Miguel Candel (Barcelona: Labor, 1973) (Original French edition, *Le catholicisme entre Luther et Voltaire* [Paris: Presses universitaires de France, 1971]); Michel Vovelle, *Piété baroque et déchristianisation en Provence au XVIIIe siècle: les attitudes devant la mort d'après les clauses des testaments* (Paris: Librairie Plon, 1973).

[4]　José Ignacio Ruiz Rodríguez and Igor Sosa Mayor, "El concepto de la 'confesionalización' en el marco de la historiografía germana," *Studia Histórica. Historia moderna* 29 (2007): 279–305; Heinz Schilling, "La confesión y la identidad política en la Europa de comienzos de la Edad Moderna (ss. XV–XVIII)," *Concilium* 6 (1995): 943–55; R. Po-Chia Hsia, *Social Discipline in the Reformation: Central Europe, 1550–1750* (London: Routledge, 1989); R. Po-Chia Hsia, *The World of Catholic Renewal, 1540–1770* (Cambridge: Cambridge University Press, 1998); Paolo Prodi, ed., *Disciplina dell'anima, disciplina del corpo e disciplina della società tra medioevo ed età moderna* (Bologna: Il Mulino, 1994); Adriano Prosperi, *Tribunali della coscienza. Inquisitori, confessonari, missionari* (Turin: Einaudi, 1996); Federico Palomo, "'Disciplina christiana'. Apuntes historiográficos en torno a la disciplina y el disciplinamiento social como categorías de la historia religiosa de la alta edad moderna," *Cuadernos de Historia Moderna* 18 (1997): 119–36.

confessional authorities and agents in intermediate levels of the social pyramid (families, fraternities, guilds, and others); and, above all, they argue against the rigidity and hermeticism of this interpretative paradigm. This chapter builds on these critiques and argues that it is necessary to present the flexibility and porosity of negotiations and adaptation in the everyday life of men and women during the Old Regime.[5] Using examples of everyday magic from early modern Spain, especially Catalonia, it demonstrates that we must look at European religious history not only as a confrontation between paganism and Christianity, but also as a more or less negotiated synthesis between theoretical norms and transgressions, a synthesis that permeated all socio-cultural niches.[6]

Negotiating Theories of Superstition and Everyday Magic

In early modern Catalonia local communities adapted norms through a form of cultural negotiation that allowed them to reorganize priorities in the light of their own traditions, while always believing that they remained orthodox. This is the "local religiosity" William Christian wrote about when discussing Philip II's Spain. For Christian, all popular religious expressions, regardless of how heterodox they would have seemed to the authorities that carried out reforms, were part of the everyday Christian beliefs of Old Regime communities.[7] Those who had to "impose" new practices and rites, like parish priests, missionary friars, preachers, and learned elites, usually tolerated this negotiation and its products: the devotions, practices, and rites that did not adhere strictly to orthodoxy. Many times, they themselves proposed and were actively engaged in this process of mystification. With regard to Catalonian history, Henry Kamen and Martí Gelabertó have provided excellent contributions using this viewpoint.[8]

---

[5]   Ute Lotz-Heumann, "The Concept of 'Confessionalization': A Historiographical Paradigm in Dispute," *Memoria y civilización* 4 (2001): 93–114.

[6]   Excellent examples of this perspective are found in Manuel Peña Díaz, ed., *La vida cotidiana en el Mundo Hispánico (siglos XVI-XVIII)* (Madrid: Abada Editores, 2012).

[7]   William Christian, Jr, *Religiosidad local en la España de Felipe II*, trans. Javier Calzada and José Luis Gil Aristu (Madrid: Nerea, 1991), 13–14 (Original English edition: *Local Religion in Sixteenth Century Spain* [Princeton: Princeton University Press, 1981]). While using the concept of "popular religion," Agustin Redondo also expresses the difficulty in defining it: "La religión populaire espagnole au XVI siècle: un terrain d'affrontement?" in *Culturas populares: divergencias, diferencias conflictos* (Madrid: Casa de Velázquez, 1985), 329–69.

[8]   Henry Kamen, *The Phoenix and the Flame: Catalonia and the Counter Reformation* (New Haven: Yale University Press, 1993); Martí Gelabertó, "La palabra del predicador. Contrarreforma y superstición en Cataluña (siglos XVII–XVIII)" (PhD diss., Universitat Autònoma de Barcelona, 2003). Although this thesis was published in 2005 (Lleida: Milenio),

Henry Kamen has emphasized the significance of the rural community in Catalonia between the sixteenth and eighteenth centuries by contrasting it to the only considerable urban center in the region, the city of Barcelona. With more than 85 percent of the population living in the countryside, a sparse habitat, and a rugged natural environment in the majority of the principality (one-third of its geography is mountainous due to the Pyrenees), daily life for the majority of Catalans was tied to experiences in the local community. These revolved around the parish as a social, economic, political, and religious axis of identity. The local religiosity of these communities was reflected in devotional practices rather than in knowledge of Christian doctrine or individual religious reflection, and this religiosity was pragmatic and affective in nature. It was linked to a process of sacralization of the land that started in the fifteenth century and had filled the landscape with hermitages and small churches high up in the mountains: places made sacred by miraculous apparitions which in most cases appealed for moral and penitential reform to overcome crises and epidemics.[9]

It was not easy to introduce and consolidate the reforms of the Council of Trent and even less easy in early modern Catalonia, which was populated by a conglomeration of rural communities with strong local identities. Reform of worship met with various forms of resistance. The improvement of priests' training, in general, and the cure of souls, in particular, was slow and quite deficient, clashing with the benefice system that favored local priests who were born in the community. For this reason many of the priests were closer to the popular culture of their surroundings than to the guidelines of Trent. They were also more susceptible to local pressure than orders coming from their diocesan leaders. The interior missions that religious orders initiated beginning the latter part of the sixteenth century had as one of their goals the re-conversion of the Catalan peasants, who resided in an interior "Indies" that had now been rediscovered as missionary territory. The religion of the majority of Catalans continued to be simultaneously Catholic and natural without ceasing to be Christian.[10]

To substantiate this assertion fully, we would have to address a vast subject area that the church judged as "superstitious," a label whose semantic content

---

I have cited from the original, which is at http://tdx.cat/bitstream/handle/10803/4793/mgv1de1.pdf?sequence=1 (accessed 2 Oct. 2012). See also Ricardo García Cárcel, *Historia de Cataluña*, vol. 1 (Barcelona: Ariel, 1985), 396–420. For Castile, there is also the excellent book by Sara T. Nalle, *God in la Mancha: Religious Reform and The People of Cuenca, 1500–1650* (Baltimore: Johns Hopkins University Press, 1993).

[9]   William Christian, Jr, *Apparitions in Late Medieval and Renaissance Spain* (Princeton: Princeton University Press, 1981).

[10]   Ignasi Fernández Terricabras, "The implementation of the Counter-Reformation in Catalan-Speaking Lands (1563–1700): a successful process?" *Catalan Historical Review* 4 (2011): 83–100.

would include different forms, such as witchcraft, sorcery, and non-orthodox uses of and practices in Catholic devotions and rites. In this chapter, the focus is on practices and beliefs that were at the heart of the laity's religious life—masses, prayers, and the veneration of saints—and those commonly associated with and even linking magic and witchcraft. Generally, the church would admit that all forms of superstition could potentially have some relationship, implicit or explicit, with the devil. In particular, witchcraft (*brujería*) implied a pact with the devil, whereas sorcery (*hechicería*) did not. Magical practices received very different consideration in society depending on whether they caused damage (an evil eye that harmed people or animals, for example) or people realized that they were carried out for the community's benefit (healings, love potions, recovering lost objects, finding treasure). Sorcery or magic could be *white*, meaning that it was beneficial for the individuals and the community, or it could be *black* and, hence, closer to witchcraft. Healers, who had different titles according to their "specialities," could cross the bounds of superstition with extraordinary ease. Also, alchemists, astrologers, and necromancers were able to enter this ambiguous territory, in which it was acceptable to read the stars but not to practice divination. Nevertheless, all these practices fell within the scope of heresy through a process of codification that began in the fourteenth century and culminated in the bull of Sixtus V, *Coeli et terrae* (1585), which included all esoteric expressions of the heretical condition and spread this information more generally.[11]

Such superstitions appealed to the supernatural through means the church did not endorse, and they revealed beliefs that did not reflect Catholic dogma. Preachers and theologians tended not to distinguish between various types of condemned superstitious practices; rather they saw all as interrelated in a complex mixture of religious ignorance and diabolic influence. The missionary Antonio Escobar exemplified this attitude when he wrote in 1645, "Superstitions tacitly invoke the devil, because they ask for the means to do things and do not have virtue for such effects. Superstition is a religion against the true God."[12] Preachers, missionaries, confessors, and inquisitors worked to show errors, correct them, punish these practices, and internalize orthodox

---

[11] Juan Blázquez Miguel, *Eros y Tánatos. Brujería, hechicería y superstición en España* (Toledo: Editorial Arcano, 1989). Regarding the origins of the concept of diabolical heresy and of the stereotypes about it, see Jeffrey Burton Russell, *Lucifer: The Devil in the Middle Ages* (Ithaca: Cornell University Press, 1984); Francisco J. Flores Arroyuelo, *El diablo en España* (Madrid: Alianza Editorial, 1985); James Amelang and María Tausiet, eds, *El Diablo en la Edad Moderna* (Madrid: Marcial Pons, 2004). For an overview of early modern European demonology, see Wolfgang Behringer, "Demonology, 1500–1660," in *The Cambridge History of Christianity. Reform and Expansión, 1500–1660*, ed. R. Po-Chia Hsia (Cambridge: Cambridge University Press, 2007), 406–24.

[12] Antonio de Escobar y Mendoza, *Examen de confesión y práctica de penitentes, en todas las materias de teología moral* (Barcelona, 1645), 18.

belief and practices. For example, confessors had to interrogate penitents within the strictures of the sacrament of penance for violations against true religion and divine worship as specified in the confessors' manuals prepared for the bishop of Barcelona, Dimas Loris, who had been deeply involved in developing and disseminating the Tridentine provisions in his diocese:

> If he [the penitent] has adored idols or other creatures ... that do not belong to God our Lord: [it is] a deadly sin. If he believes in dreams, omens, spells, sorcery, and superstitions or if he has used any of these to hurt his soul or that of his fellow man: a deadly sin. If he has given credence to or has superstitious lists with figures, signs, and names of demons, or of obscure and unknown names that are not approved of: a deadly sin. If he has cooperated in witchcraft or has had a pact with the devil to accomplish bad things (*maleficia*): a deadly sin. If he has forbidden books on these subjects or heretical [books] that haven't gone through legitimate correction and approval: a deadly sin. If he has mixed profane and divine things with irreverence and contempt for the divine: a deadly sin. If he has made devotions for a notably bad purpose, as in killing, etc.: a deadly sin.[13]

The majority of the questions fell into the category of a generic mixture of profane and divine things, a confused combination that was equated with superstition in the eyes of the Counter Reformation Church.

To these debates on superstition was added, in the sixteenth century, an outpouring of demonological studies with texts by authors such as James I of England (1579), Jean Bodin (1580), Nicolas Rémy (1585), Peter Binsfeld (1589), Martin Del Río (1599), Henri Boguet (1602), Francesco Maria Guazzo (1608), and Pierre de Lancre (1612). This demonological fervor has been attributed to the resurgence of scholastic realism and Augustinianism that argued for the reality of diabolic action in both male and female witches. It has also been linked to the relative decline of neoplatonism and nominalism, two schools of thought that had been inclined to minimize the importance of the devil.[14] Whatever the source of such attitudes, in early modern Catalonia the integration of superstition and demons in influential guides made the process of negotiation more fraught and the consequences of failure more frightening.

Yet their effects on the broader community and ideas are much more debatable. If the authors of these treatises endeavored to create an exhaustive

---

13   *Memorial de manaments y advertencias del Molt Iltre. y Reverendissim señor don Joan Dymas Loris bisbe de Barcelona y del Consell de Sa Majestat & per als sacerdots, confesors, rectors y curats de son bisbat* (Barcelona: Gabriel Graells & Giraldo Dotil, 1598), fols. 57v–58r. The manual's author was the Jesuit Pere Gil, who played a very important role in witchcraft episodes that were more significant than those that occurred in Catalonia, as we will see in the following pages.

14   Russell, *Lucifer*, 311–12.

taxonomy of different forms of superstition, in popular culture such conceptual distinctions were not made. Appeals to the supernatural could be either orthodox or heterodox depending on the individual person making them and their spatial and chronological context. Their assessment by society depended on whether the result harmed the community or benefited individuals or the group. This was pure functionalism—as functional as was resorting to witch hunts as a violent escape from social pressure at certain moments. It is important, however, to recognize adequately the scale of these occurrences. We must distinguish between the witch hunt and the very different dynamics of that which we would call "everyday magic." Although affecting a large number of victims, the witch hunt did not represent the typical experience of the majority of the population. They experienced what we could call "normal" witchcraft; it focused more on curses and sorcery and less on the mythology of the witches' sabbat and satanic covens.[15] In the following pages we will address some of these forms of recourse to the supernatural in early modern Catalonia—from the episodic instances of sorcery to the endemic presence of more or less authorized magical practices.

Catalonia's Witch Hunt and the Evolution of Learned Discourses

According to inquisitorial sources, the witch hunt in Spain was a minor one. Between 1550 and 1700 there were 3,532 persons prosecuted for superstition, including witchcraft, sorcery, and all esoteric variations of magical practices, a figure that represents 7.9 percent of all those prosecuted during this period.[16] In Catalonia the total of those prosecuted for superstition was 419 between 1598 and 1820, which comes to 7 percent. The number of death sentences in Spain was also very low, around 1 percent.[17]

The attitude of the Inquisition and some of the clergy with regard to sorcery was skeptical, but the power of the devil was beyond doubt. Even that power was subject to the divine will and purpose, however, a perspective that followed medieval precedent. A good example of this dynamic was the biblical story of Job. The problem the inquisitor faced had to do with the instruments available for a judge to prove witchcraft. It is exactly here where, guided by profound misogyny, inquisitors and theologians faced what appeared to be an

---

[15] James Amelang, "Invitación al aquelarre: ¿hacia dónde va la historia de la brujería?" *Siglo de Oro* XXVII (2008): 29–45; Wilson, *The Magical Universe.*

[16] Jaime Contreras and Gustav Henningsen, "Forty-four thousand cases of the Spanish Inquisition (1540–1700): Analysis of a historical data bank," in *The Inquisition in Early Modern Europe: Studies on Sources and Methods,* ed. Gustav Henningsen (De Kalb, IL: Northern Illinois University Press, 1986), 100–129.

[17] Joan Bada, *La Inquisició a Catalunya (segles XIII–XIX)* (Barcelona: Barcanova, 1992), 133–36.

insurmountable barrier. Taking into account feminine fragility, could women's testimony be accepted? If women could so easily be deceived by the devil, and the judges' suggestions as well as fear of torture could so seriously affect them, how could their testimony be valid?

Since the Congregation of Granada in 1526, which had gathered precisely to discuss this matter, the Inquisition affirmed that cases of witchcraft were to be handled with care, and it had to be taken into account that the most likely explanation was that these events were most probably a demonic illusion instilled in female minds. The most efficient solution was then education as well as discretion in the treatment of the accused. Although the Suprema (Supreme Council of the Inquisition) assumed this rule, it was much more difficult for inquisitors in the courts to implement it due to pressure from local communities. In one example from Catalonia, a local inquisitor burned six women at the stake between 1548 and 1550 before a specialist inquisitor arrived to bring order, liberate the women who were still imprisoned, and return honor to the women who had been killed. The personnel of the tribunal were severely reprimanded, and the witch hunter himself was burnt in a public auto-da-fé.[18] Despite such cases, overall the Inquisition was seen as a more benevolent court than the local courts of justice. For example, in 1575 a local court condemned Margarita Boer to the gallows, but she managed to save her life by getting her case into the hands of the Holy Office.[19]

Especially serious was the brutal witch hunt of 1611–1627, which also cut through central Europe.[20] It is difficult to provide a precise number of victims in Catalonia because so many sources have been lost, but recent estimates suggest that 600 women were hanged. The persecution took place mainly in the valleys of the Pyrenees and central Catalonia and later in the regions of Roussillon and Cerdanya. Local civil authorities carried out trials against witches without inquisitorial intervention. As Gunnar Knutsen has demonstrated, the unleashed witch hunt correlated with a rebalancing of power and status between local elites.[21]

---

[18]   For witchcraft and the Inquisition in the territories of the Crown of Aragon, see William Monter, *Frontiers of Heresy: The Spanish Inquisition from the Basque Lands to Sicily* (Cambridge: Cambridge University Press, 1990), 301–22. For the witchcraft trials of 1548–1550, see Doris Moreno, "Las estrategias inquisitoriales ante la brujería en la Cataluña de 1548," *Una historia abierta. Homenaje al profesor Nazario González* (Barcelona: Universidad de Barcelona, 1998), 39–47; Pau Castell i Granados, *Un judici a la terra dels bruixots. La cacera de bruilles a la Vall Fosca, 1548–49* (Tremp: Garsineu Edicions, 2011).

[19]   Archivo Histórico Nacional [hereafter AHN], Inquisición, libro 730, fol. 193.

[20]   Antoni Pladevall, "La cacera de bruixes a Catalunya": http://bruixes.mhcat.net/images/stories/pdfs/article_07.pdf (accessed 27 Dec. 2012). A complete bibliography is available at http://bruixes.mhcat.net/index.php/ca/per-saber-mes.

[21]   Gunnar W. Knutsen, *Servants of Satan and Master of Demons. The Spanish Inquisition's Trials for Superstition, Valencia and Barcelona, 1478–1700* (Turnhout: Brepols, 2009), 85–115.

In the majority of cases the trial was brief and implementation of the verdict was immediate because of escalating popular pressure. Many women who were sentenced to death by hanging betrayed each other and, in most cases under torture, confessed to stereotypical practices of which many were taken from the demonological manuals that circulated as well as from sermons and popular rumors. These practices included nightly travels, transformation into animals, preparation and application of ointments, attendance at sabbats, and submission to the devil who appeared as a goat.[22]

The complexity of the crime, the intervention of various jurisdictions, the extent of the affected area, the significant social fear provoked by the phenomenon, and the brutal procedures of local authorities alarmed the upper political classes. One of the most prominent theologians in Barcelona during those years was the Jesuit Pere Gil. He was a consultant to the Inquisition in Barcelona from 1604 until his death in 1622, and he was very close to the circles of power in Catalonia. In 1619 Gil wrote a memorandum on the problem of witches, a memorandum that the Suprema as well as the Viceroy of Catalonia received. It included a proposal for how the Holy Office could intervene more effectively in such cases. Gil proposed that the Inquisitor General should request exclusive jurisdiction over witchcraft from the Pope because only that "would help protect the innocent."[23] In order to calm the situation he suggested that an inquisitor with extensive discretionary powers visit the affected areas to "improve the souls that today are blind and without light and who, for lack of faith have fallen," and he should also provide absolution and gentle forgiveness.[24] Philip II took Gil's proposal and consulted with the Viceroy Duke of Alcalá about it some months later. As for the memorandum, Pere Gil highlighted the procedural irregularities from which the alleged witches suffered, and he questioned the supposed indications of their crimes. Lastly, he called for caution and gentleness in punishment.[25]

The debate continued as the stream of condemned women kept flowing until at least 1627. To that end it is worth paying attention to the first edition in Catalonia (1628) of *Tratado en el qual se reprueban todas las supesticiones y hechizerías: muy útil y necesario a todos los buenos Christianos zelosos de su salvación* (*A Treatise Reproving All Superstitions and Forms of Witchcraft: Very Necessary and Useful for All Good Christians Zealous for Their Salvation*) by Pedro Ciruelo

---

22   Russell, *Lucifer*, 312.

23   Biblioteca Nacional de España [hereafter BNE], ms. 2440, fols. 85r–88r.

24   BNE, ms. 2440, fols. 85r–88r. The document is unsigned but is in the style and tone of the Jesuit Pere Gil. The memorial is at fol. 89r–93r.

25   Agustí Alcoberro Pericay, "Los otros 'abogados de las brujas'. El debate sobre la caza de brujas en Cataluña," in *Akelarre: la caza de brujas en el Irineo (siglos XIII–XIX). Homenaje al profesor Gustav Henningsen*, ed. Jesús Ma. Usunáriz (Donostia: Eusko Ikankutza, 2012), 92–115.

(1470–1548). The book was originally printed in 1530, and ten more editions appeared during the sixteenth century. A professor of theology at the University of Alcalá and, above all, a mathematician and an astrologer, Ciruelo stood out in the sixteenth century for the punctilious and pointed orthodoxy that he used to criticize the reputation of Erasmus in the famous Congregation of Valladolid of 1527, which judged the Dutch humanist's works.[26] The publication of the new, seventeenth-century edition of Ciruelo's *Treatise* was carried out by order of the Viceroy of Catalonia, Miguel Santos de San Pedro, who was also bishop of the diocese of Solsona, one of the areas affected by the witchcraft epidemic. It was published with some additions by Antoni Jofreu, a prominent jurist who would have a significant role in the following decade defending constitutionalism against the claims of the Count-Duke of Olivares.

In the *Treatise*, Ciruelo strongly supported the reality of witches' testimonies and advocated punishment in keeping with the laws. The novelty of the 1628 edition was based on Jofreu's additions in which he softened Ciruelo's original stance. In the foreword Jofreu praised the treatise's extraordinary usefulness, reaffirmed the identification of witchcraft and sorcery with heresy, and demanded capital punishment for those convicted of these crimes, referring to St. Augustine, his attitude to the Donatists, and interestingly, the example of Calvin: "Even the pervert Calvin rendered justice to Miguel Serveto for being a heretic: ([that is], that superstitious people end up as heretics) and he, together with his disciple Beza, wrote that heretics should be punished."[27] Jofreu also updated Ciruelo's treatise by adding a Tridentine vision on superstitions in worship and in piety by referring to authors like Martín Del Río, Torreblanca Villalpando, or Prospero Farinacci.

Jofreu agreed completely with Ciruelo about the standard rules, but he was less firm when it came to practice. For example, he expressed a more juridical viewpoint when he differed from Ciruelo in his skepticism regarding the nature of the crime: "I would point out the countless positive acts and perfidies that these women declare are false and, thereby, apparent and superstitious. Not only do they deceive themselves but also learned people and holy men."[28] In his analysis of superstitious prayers, Jofreu expanded the area of "permitted things" whereas Ciruelo was very strict and warned about how wrongly "good and saintly things" could be used. The most common errors people made in saying prayers and psalms were (1) to use them to ask for things contrary to the ones God had ordained; (2) to pray using irregular forms, such as lies or falsehoods

---

[26]   Alva V. Ebersole, "Pedro Ciruelo y su *Reprobación de hechicerías*," *Nueva Revista Filológica Hispánica* 16 (1962): 430–37.

[27]   Pedro Ciruelo, *Tratado en el qual se reprueban todas las supersticiones y hechizerias: muy útil y necesario a todos los buenos Christianos zelosos de su salvación* (Barcelona: Sebastián Cormellas, 1628), paragraph 10 of the "Prólogo primero."

[28]   Ibid., 53.

against the faith or natural reason, and "unknown and barbaric names, and other mockeries and frivolities"; and (3) to give the ceremonies' accompanying prayers a thaumaturgical emphasis. In Jofreu's comments, he introduced interesting nuances and modulations, probably because of the popularity of many of these practices. In his opinion psalms and prayers that were not mixed with clear superstitions or unknown words were not *"absolutely illicit"* [emphasis added]. Jofreu argued that, for instance, in Spain many healers were allowed to heal if they had been approved and tested "and use[d] their craft to heal, even if they were wicked men, because they had this craft through grace."[29] For Jofreu, the ability to heal could be a supernatural grace. As such, he held to the orthodox interpretation against Donatism: even despicable men could be granted the ability to heal. The practice of a divine grace by a contemptible person, with the approval of ecclesiastical authorities, allows Jofreu to emphasize that very clear evidence of superstition distinguishes that which is absolutely illicit. Where there was no clear finding, the jurist could allow healing practices to continue.

For Ciruelo before the Council of Trent as for Jofreu after the Council, witchcraft and sorcery were one and the same and they both led to heresy. But to establish the limits between them was not easy; to distinguish the *"absolutely illicit"* from the generally licit, as Jofreu did, left a vast area in between where many practices could take place. As their example shows, the learned elite in early modern Catalonia had difficulties in purging the illicit from the vast world of the supernatural in the realm of everyday life.

Witches and Demons for Everyday Life

Detached from pacts with the devil and polarized with what we call sorcery, witchcraft was already recorded in documents from medieval Catalonia. The first edict against witches who presumably participated in the sabbat was enacted in the Vall d'Aneu in 1424, but as far as we know, this evidence of witchcraft was episodic in nature. More commonly, male and female sorcerers were regular figures in the social landscape of late medieval Catalonia, and they caused unbearable fear only in certain moments. Some locals were considered witches in popular opinion for many years without necessarily being formally accused of it.[30] Reports by diocesan visitors from the sixteenth to the eighteenth centuries, however, show that social unrest in the daily life of many rural communities provided a fertile ground for the emergence of phenomena related to witchcraft.

---

[29] Ibid., 229–30.

[30] In 1619, the witnesses in the trial against Antoni Pons, accused of witchcraft, testified that he had been rumored to be a witch for at least ten years: Biblioteca de Catalunya, Bat. Cat. (Batllia de Cardona) 35/14, s. f.

In this context witches became an easy target, an alleged source of evil, and an exhaust valve for aggression and social anxiety.

The devil and witches in a strict sense were not present in the manuals guiding the religious orders' missions in rural Catalonia. These orders were much more preoccupied with morality, sexual conduct, sacramental life, and peace between neighbors. Demonological discourse seemed to be reserved for the clergy's exclusive use, and it was disseminated to them through manuals of Catholic doctrine written specifically for the clergy. Sermons for the laity more generally focused on combating superstitions, on teaching the difference between the profane and holy, and on determining licit access to the supernatural.

Despite tensions, or perhaps because of them, the presence of "magical personages" with supernatural powers was a constant in early modern Catalonia. They were accepted, sought after, consulted, and feared. As previously mentioned, women could easily be suspected of magical practices because of the alleged weakness of their character and/or their role in the community. Many of them accumulated experiences and knowledge transmitted from mother to daughter, however, and when widowhood or old age began to loom on the horizon they could survive by acting as midwives or traditional healers in the villages.[31] Such roles often involved some form of magic.

Practitioners of sorcery and magic were often those men or women who might be accused of "casting the evil eye." To be labeled such a person, it was sufficient to differ somehow from the norm: to be old, poor, a redhead, a tramp, a blind person, or a foreigner, among other distinctive marks. Preachers never became tired of criticizing such habitual and unfounded accusations: "A woman could have a baby who's ill, or someone would hurt the baby on the head. And because someone has seen them before, now the only thing they talk about is that this person has cast the evil eye on the child—he looked at me this way, now I am bewitched. And they say it with such conviction that no one can make them believe otherwise."[32] Another frequent accusation could be that the person caused impotence in men or infertility in women. The procedure to make a man impotent consisted of the sorceress making three knots in the drawstring of a pair of pants and exiting through a door while saying a short prayer. The sorcery was done at the behest of others, and once it was completed, the pants were returned secretly to the owner. Word spread that, to prevent this from happening, the engaged couple had to live together in the same house before getting married. A synodal constitution of Lleida in 1691 warned that such procedure did not

---

[31]   Robert Muchembled, *La sorcière au village (XVe–XVIIIe siècles)* (Paris: Julliard, 1979).

[32]   Biblioteca Universitaria de Barcelona [hereafter BUB], Anonymous, *Explicació dels manaments*, ms. 1424, fol. 154.

make any sense and, above all, it was contrary to the decrees of the Council of Trent.

Individuals who possessed magical powers were at once attractive and frightening. They were intermediaries for a magical world where one could find answers to many kinds of problems. In one case, for example, during 1561, when the inquisitor visited the region of Vic, many magical practices such as spells, divination, and the summoning of storms were reported after a public reading of the Edict of Faith. Caterina Reig, a 68-year-old widow, was accused of conjuring in order to find missing people. She did this by thrusting a pair of scissors into the ground; on it she placed a sieve, and then she recited, "For St. Peter and St. Paul, sieve, for the virtue which God has given to you, if Francina is in ... [here she recited names of places] ... turn if yes, and if not, don't move."[33] People also had to protect themselves from those individuals who could use their magical power against others. The resources for this protection were basic and at everyone's disposal. Although the church had specified the devil's powers and insisted that these powers were real, at the same time its representatives criticized superstitious practices that were related to popular beliefs that, in a magical world, people could routinely use to exorcize evil powers. The theologian Pere Salses, who knew rural Catalonia very well due to the years he spent as rector in the parishes of the Urgell diocese, expressed this tension very clearly in the mid-eighteenth century: "[A]ll the natural remedies used against witches are superstitious. The broomstick behind the door, the egg shells, the sprinkled salt, the needles, they are all futile and superstitious remedies, and to use them is to invoke the devil, when you wish to pull away from him, and it is a mortal sin of which you can only be excused because of ignorance."[34]

The presence of healers who claimed to be witch hunters, such as Joan Malet the Morisco (1548–49) in Tarragona or the Frenchmen Tarragó and Carmell (1617–20) in northern Catalonia, stimulated confrontations with and persecutions of witches. Local authorities quite frequently gave these witch hunters resources before communities articulated rumors about witches and experienced inexplicable phenomena. Such hunters were able to make cohabitation in these communities unbearable while presenting themselves as an optimal resource in different situations: as instruments of personal vengeance in case of grudges between families or neighbors or as providers of plausible explanations for illnesses or harmful and unexpected atmospheric phenomena. The church tried to mediate in these cases although with little success. In 1749,

---

[33] Cited by Antoni Pladevall i Font, *Persecució de bruixes a les comarques de Vic a principis del segle XVII* (Barcelona: Els Comtes de la Vall de Marlès, 1974), 35.

[34] Pere Salses, *Promptuari moral sagrat, y cathecisme pastoral de platicas doctrinals y espirituals sobre tots los puntos de la Doctrina Christiana, per predicar en la Quaresma, Diumenges, y demes festivitats, Rogativas per aygua, y altres necessitats, ab índices de la Sagrada Escriptura, de las Platicas y cosas mes notables que en se contenen* (Barcelona, 1754), 135–6.

the rector of a parish in Vic denounced a blind old man from his community, Ramón Closa, to the Holy Office; although for many years Closa had been known as a healer, now he only caused alarm because he claimed that many people who had recently gotten sick were actually bewitched.[35] In 1772 the episcopal visitor from the Vic diocese warned that such individuals provoked "dissension in the villages because of suspicions, rash judgments, and gossip [that they incite], suspecting each other of bewitching them or of bringing about the ills they are suffering, or think they are [suffering], according to their apprehension." He called for strong action by the rectors to report these cases to judicial authorities.[36]

The devil had his rightful place when it came to possessions and exorcism or to natural disasters. The medieval church was directly involved in controlling nature, especially when demons appeared to guide it. A liturgical manuscript written at the end of the fifteenth century, which covers the ceremonial practices in the bishopric of Vic, gives a partial description of an exorcism that a priest must carry out against aerial demons who caused terrible tempests of rain, hail, and lightening. To conjure atmospheric demonic armies, the priest had to put on the sacred habit of the High Mass and hang the stole and cross on his breast. Then he had to find a place in the open air, draw a large cross on the ground, and write between the arms of the cross, "jesus nazarenus=sidrach=misach=et Abdonage." A series of prayers and curses followed that were designed to repel evil spirits from the air. This rite was later purged from liturgical texts printed by the dioceses, but in the following centuries people continued to conjure against storms or invasions of locusts and rats.

Familiar spirits were among the other demonic and invisible characters that inhabited supernatural space in Catalonia. Some of the witches persecuted in the seventeenth century confessed to having a familiar spirit who helped them. Probably the most famous incident of a familiar was the one that took Dr. Torralba on a nightly visit to contemplate the Sack of Rome in 1527.[37] Some of those brought before the Inquisition confessed to having a benevolent familiar with whom they clarified scientific doubts. Contrary to the opinion of those who benefitted from such spirits, the Inquisition always considered them an evil force. When the chief Inquisitor Alonso de Manrique reformed the edict on denunciations in the beginning of the 1530s he introduced various chapters on magical arts, divination, and astrology, and in one of them he asked for the prosecution of those who claimed to have familiar spirits. Although more

---

[35] Van Pelt-Dietrich Library Centre (University of Pennsylvania), Thorndike Collection of Inquisition Manuscripts [hereafter Thorndike Collection], Mss. Coll. 49, exp. 10.

[36] Archivo Episcopal de Vic, Visitas pastorales, libro 1224, fol. 106, cited by Gelabertó, "Palabra del Predicador," 151.

[37] Caro Baroja, *Vidas mágicas*, 1:231–94.

common during the Renaissance, the reputation of familiar spirits continued over time. In 1749 the blind man Ramón Closa presumed he had a familiar spirit when he was a soldier and knew that other people had had one, although later he did not dare confess it.[38]

Exorcisms of all kinds of spirits were applied frequently to the sick, in secret and in private homes, often mixing the official liturgy with heterodox and magical practices. People appreciated exorcistic practices because, in addition to liberating the sick person's body from its suffering—suffering that diabolic intervention had caused—its baroque theatricality moved the assistants. Clerical authorities mistrusted such practices, even while providing details about them. The diocesan synod of Lérida in 1716 complained that "some clergymen use private books of spells that do not have the necessary approbation and include various superstitions as well as words that are not at all true to Our Faith, and, similarly, because of the credit ordinary people give these actions, [the clergy use them, and] their fervor, the shouting, and the ridiculous gestures undermine the gravity of the ministry and give cause to rumors and gossip."[39]

Exorcists and parish priests could act in the genuine interest of their communities' members, but sources also note the clear economic benefits they received as they took advantage of their authoritative positions. In 1639 15 witnesses declared that Bernat Rafael, beneficed clergyman of the parish of Santa María del Mar of Barcelona,

> cures spells with prayers and holy water by putting on the stole, and later when he sees them he says that they are enchanted or bewitched, and he takes money for these cases ... and having made the daughter sit on a chair, he put a stole on her neck and part of her body and he tied in one of her arms, and having left the sick woman like this he put on the blessed light and he read from a book and he made the sign of the cross on her and after having read for a while he took a Host which had letters printed on it in ink and he gave that to the sick woman to eat saying when giving it to her, in name of the Father, the Son and the Holy Spirit, and he said to her three Lord's Prayers and three Hail Marys, and then he blessed the water and gave it to her to drink after the Host and he also blessed the bread and water which the sick woman had brought with her and he told her to come back for eight consecutive days and not to drink anything other than blessed water and eat other than blessed bread.[40]

---

[38]   Thorndike Collection, Mss. Coll. 49, exp. 10.

[39]   Archivo de la Catedral de Lleida, Sínodo de Lérida (1716), tit. II, Const. VII, fol. 45, cited by Gelabertó, "Palabra del Predicador," 293.

[40]   AHN, Inquisición, lib. 734, fols. 168–9.

The inquisitors reprimanded him severely, and he was forbidden to carry out exorcisms using these procedures.

The misuse of sacred objects was not exclusive to priests. Some bishops had a more open albeit traditional approach to the liturgical tradition prior to Trent, and they allowed the laity to use sacred elements and objects for their own benefit as long as they did not exceed certain limits when applying them. From the first half of the sixteenth century, Catalan reformers condemned categorically the use of liturgical elements to perform superstitious cures. Diocesan visitors continually warned parish priests to be especially cautious about monitoring holy water in the baptismal font closely, but to no avail; for example, the diocese of Vic tolerated the use of holy water as a domestic means of protection. Other dioceses also accepted uses of holy water that clerical reformers saw as more questionable. The episcopal hierarchy of the diocese of Urgell also encouraged the laity's use of holy water as a marvelous talisman to neutralize perils like the presence of demonic spirits within houses or people or to prevent invasion by locusts, rats, or other pests.

In the Face of Disease

Healers were commonly present and accepted by the general population; they even had distinctive names according to their "technical specialties": if they cured with spit, divination, prayers, or signs; if they cured the evil eye or other enchantments, dislocated limbs, etc.[41] They were, almost invariably, seen as a trustworthy resource and perhaps the only one available for people facing difficulties. From Catholic rituals healers borrowed signs and prayers that enhanced their magical powers to cure diseases. The secrecy involved reinforced the mystery that surrounded their implementation:

> Healers use prayers, crucifixes, signs, and other useless and disproportionate methods to achieve health that are not instituted by God or by the Church, whether they have the virtue to heal [or not]. ... he has a heart disease or epilepsy, and in order to bring him round they whisper into his ear certain words in secret, and they say this is sufficient for him to become better. And if the words are not said when requested and in secret they [the healers] say they would not have any effect.[42]

---

[41]   *saludadors, endevinaires, xucladors, oracioners, senyadors, trencadors, setens, desagulladors, samaires*: for a detailed description of the curative specialties of these healers, see Joan Amades, *Bruixes i bruixots* (Tarragona: Edicions el Mèdol, 2002; 1st ed., Barcelona, 1934), 41–7.

[42]   BUB, *Explicació dels manaments*, fol. 126.

Even among such opponents, magical healing did not imply by definition a pact with the devil. The ecclesiastical authorities of Catalonia attributed the existence of healers, who were easy prey for the devil's tricks and illusions, to ignorance and not to a premeditated search to communicate directly with demonic forces in order to obtain benefits and augment the thaumaturgical powers with which the common people usually associated healers. The leniency of the sentences the Inquisition imposed on this type of activity shows how mild the repression was: the most common sentence was a rebuke, pilgrimage to Montserrat, recitation of some prayers, and financial penalties.

The Counter Reformation Church fought these practices with pedagogical procedures that intended to show their complete uselessness. Confessors had a key role in this battle. The Catalan church of the seventeenth and eighteenth centuries faithfully followed the instructions of the bishop of Vic in the late sixteenth century, Benet de Tocco. He called on his confessors to counsel penitents wisely and skillfully on the use of healers and not to absolve them without being sure that they would never use one again "because they regularly have more faith in this nonsense than in God or the Roman Church."[43] From the pulpits preachers remarked on the same aspects, as did an anonymous, seventeenth-century missionary: "healers, who wander in the markets, taverns and the inns, are nothing more than money grubbing people and they don't have any virtue of healing."[44] There were, however, "authorized" healers to whom the church granted special grace and permission so that they could perform their craft. In December 1618 Pere Català obtained permission to exercise the therapeutic practices of a healer in the bishopric of Vic. On 17 May 1668 the bishopric of Barcelona authorized Jaume Puig, a resident of Manresa, to come to the capital city of Barcelona and, through his curative techniques, alleviate the disastrous effects caused by rabies in people and animals.[45]

An exceptional case was that of French healer Bernat Rigaldía, on whom the city of Barcelona called at a critical moment during 1589 when the plague took the life of 2,000 people in only six months. The despair of the city's authorities led them to hire this healer who was renowned for his efficiency in handling disasters such as the one devastating the city. Upon arriving in Barcelona, Rigaldía ordered a Novena of masses to St. Blaise and a general rogation and penitential procession. He also demanded a hefty remuneration for himself and his assistants. The citizens of Barcelona followed him through the streets in great numbers, while the most powerful people hired him privately to care for their

43 Cited by Gelabertó, "Palabra del Predicador," 334.

44 Archivo de la Corona de Aragón [hereafter ACA], *Sermon missional contra la superstición*, Monacales-Universidad, leg. 130, np, cited by Gelabertó, "Palabra del Predicador," 345.

45 Archivo Diocesano de Barcelona B, *Registra Gratiarum*, vol. 68, sn, cited by Gelabertó, "Palabra del Predicador," 345.

families. People started to talk about miracles. Barely a month after his arrival in Barcelona, however, the number of deceased kept rising, and the authorities began to doubt their investment. The poor felt distraught, and official doctors started to see the healer as an intolerable professional intruder. Rigaldía went from being venerated to being suspected of sorcery. It was said that he performed sorcerous practices—even more, that he himself had poisoned the citizens by scattering venomous powders on clothes and baptismal fonts. Some witnesses reported having seen a demon in the form of a calf at the Frenchman's house. Bernat Rigaldía was thus sentenced to a death designed to provide an example to others: standing in a wagon that went around the city, he had his hands cut off; then he was beheaded and cut into quarters. His head was hung on the wall of the Town Hall.[46]

Saints and Relics

The Council of Trent asserted the role of saints and their cult as intercessors for sinners before God. It ordered the education of the faithful about the intercession and invocation of the saints, the honor due to relics, and the legitimate use of images to purify worship and eliminate the superstitious practices Protestants had targeted. The cult of the saints and the veneration of relics underwent a spectacular promotion in Catholic Europe during the early modern period.[47]

In Old Regime Catalonia, the cult of the saints had two axes. On the one hand, there was the official cult with its theoretical corpus of rites and ceremonies, geared towards the veneration of the saints and common to the Catholic Christian world. On the other hand, there was the local cult—urban or rural—supported by demands for protection directed to specialist, local, or regional saints and common. In this second area devotion to the saints integrated complex components of communal identity, special devotions, and specific appeals in particular situations. In the same sense, regional and local processions or pilgrimages to and celebrations at hermitages and shrines integrated the religious life of communities by mixing religion and recreation. The baroque nature of the processions, emotional exaltation of the devotion, and elements of communal identity stirred and intensified popular passion for saints and relics.[48]

---

[46]   José Luis Betrán and Fernando Bouza, *Enanos, bufones monstruos, brujos y hechiceros* (Madrid: DeBolsillo, 2005), 183–209.

[47]   Simon Ditchfield, "Tridentine worship and the cult of saints," in *The Cambridge History of Christianity. Reform and expansion, 1500–1660*, ed. R. Po-Chia Hsia (Cambridge: Cambridge University Press, 2007), 201–24.

[48]   José Luis Betrán, "Culto y devoción en la Cataluña barroca," *Revista de historia Jerónimo Zurita* 85 (2010): 95–132.

Preachers' sermons promoted a model for veneration of saints and relics, but its imposition was neither easy nor quick. Even the learned made thorough efforts in this regard. All establishments and corporations shared the characteristics of a local religiosity that conceded vast, autonomous, and supernatural powers to the thaumaturgical saints, a status that was far from their role as mediators. The majority of therapeutic saints had a concrete curative specialty: St. Sebastian or St. Roch was invoked in cases of plague, St. Job in cases of leprosy, and St. Christopher in cases of sudden death. The church endorsed these specialties but at the same time tried to eliminate superstitious adhesions that had been incorporated into the saint's cult, popular beliefs that varied according to context and were difficult to correct due to their nature. Hence the importance of missionary preaching: missionaries could detect these variables and correct them through their sermons and the confessional.

The people also revised and expanded on the saints' thaumaturgical powers. One mechanism of elaboration consisted of drawing practical and curative consequences from a detail in a saint's life. St. Stephen, who had been stoned to death, was converted into a reliable protector against wounds and stabs, and the way the faithful used this protection could become problematic, as the Jesuit Benito Noydens explained in the seventeenth century: "It is superstitious to say or to believe that he who on the day of St. Stephen Protomartyr fasts with bread and water and confesses or communes will not die that year, even if he receives many stabs and wounds. This gives the people permission to commit many serious sins, as some who have this confidence tend to give themselves completely to vices."[49]

The curative efficiency of the saint was multiplied in popular belief with the use of images, and the diocesan synod of Barcelona in 1752 condemned the practice of putting a saint's image under a pillow to ensure healing. Papers with prayers or written phrases, sometimes unintelligible, were believed to strengthen the healing capacity of the saint. In one sermon a missionary corrected those who claimed that people could not die suddenly or be condemned if they wore the prayer Christ had dictated to his three carnal sisters one day while walking near the temple of Salomé [that is, of Solomon].[50]

Devotion towards relics grew spectacularly in this same period. It was believed that relics transmitted the saint's thaumaturgical powers. Friars who died with a reputation for sanctity and who had regularly had personal contact with poor peasants acquired similar status, and this status endured through oral transmission for a long time in people's minds. When one of these friars died, popular emotion overflowed: people wanted to seize, in any possible way,

---

[49] *Práctica de curas y confesores y doctrina para penitentes* (Madrid: Andrés García, 1679), 11–12.

[50] Cited by Gelabertó, "Palabra del Predicador," 266.

a part of their bodies or their mortuary habits as relics. These situations almost invariably led to violence that required the authorities' intervention. Upon the death of the friar Mariano de Sant Hilari, who was known for his sanctity, devout people "began with scissors and knives and in every possible manner to cut his habit, nails, [and] bangs."[51] The thaumaturgical powers of these friars were extended to their order. Missionaries sought this credit that converted them into authoritative figures with special powers when they travelled in the region.[52]

Seventeenth-century missionaries disputed with one another over territory, which was more than an area for conversion; it was a source of acquaintances and alms, of prestige and social accreditation. In this context, for example, Capuchin monks in the eighteenth century promoted devotion to and belief in powders made from the crosier of St. Francis for women in labor. The Capuchin friar Josep de Sant Celoni wrote in 1768 about the success of a cross made from St. Francis's crosier in protecting women in labor and their sons in Tarragona. He then added, "Such is the experience, the happiness this cross brings in such an episode, that they hardly let it rest in the monastery, taking it from one house to another with such frequency that on occasion much time passes without the friars knowing of its whereabouts, and if you would have written down all the remarkable cases that have happened because of it, it would fill a great volume."[53] In the eighteenth century Catalan Jesuits propagated the miraculous virtue of the water of St. Ignatius, the thaumaturgical powers of which had been acquired from having been in contact with the sacred relics of the founder of the Society of Jesus; it was especially effective when applied to women in advanced stages of pregnancy so that they would have a safe and successful delivery. It also benefitted people tormented by demons. Similarly, in their missionary campaigns the Discalced Carmelites distributed the water of St. Albert, which was considered extraordinarily effective at healing the sick bodies of those who were possessed. The Augustinian friars spread information on the incredible curative properties of prayer that occurred on and bread that was blessed on the day of San Nicholas Tolentino; the bread was particularly useful against plague and infectious diseases.

Such use of the cult of the saints took many other diverse forms that could appear similar to other, less orthodox practices. Appeals to supernatural powers, to miracles performed by the saint, could also provoke personal or communal vengeance. A saint's lack of response could lead to a violent reaction by the believer; just as the saint's supernatural intervention in the physical world

---

[51]   Cited by Ibid., 320.

[52]   Louis Châtellier, *The Religion of the Poor: Rural Missions in Europe and the Formation of Modern Catholicism c. 1500–c. 1800* (Cambridge: Cambridge University Press, 1997; 1st ed., Paris: Aubier, 1993).

[53]   Cited by Gelabertó, "Palabra del Predicador," 314–15.

was demanded, the absence of an answer could provoke a believer's physical intervention in the saint's supposedly supernatural world. Thus, the crime could enter the provenance of the Inquisition through their mandate to prevent superstition or blasphemy, as occurred in the case of Magdalena de Lleó from Balaguer, who was accused in 1666 of

> keeping under her petticoat a wooden figure of St. Anthony of Padua which, in the grief and anguish in which the accused finds herself, she whips, hangs, and burns in fire to end the grief in which the accused finds herself through the intercession of the saint. And occasionally she would say to him [the saint]: My Saint, you know that you must say a few masses and thereby get me what I ask because if you don't do it I will not only not praise you but I will also hang you and burn you in fire ... . The prisoner hanged the image of St. Anthony and she put a cord on him and hanged the image in a manger ...[54]

Masses and Prayers

While being perfectly orthodox on paper, other Catholic rituals were also highly malleable. The mass was probably one of the most important rituals as a saving act for the believer, beneficial for spiritual and physical wellbeing, but for the majority of the population, the mass had an esoteric component that made it prone to being regarded superstitiously.[55] Various Catalan liturgical manuscripts of the fifteenth and early sixteenth centuries said that the person who attended mass would not get old, nor would his body weaken while the religious service lasted. Pregnant woman were assured a good labor, and foodstuffs would be converted into physical health.[56] Throughout the Middle Ages masses were said for the most diverse aims and devotions, and pre-Tridentine liturgical books dedicate much space to describing particular rites and special masses like the one dedicated to Blessed Job against syphilis.

A considerable number of these masses were purged as superstitious from the Catholic Missal with Pius V's liturgical reform in 1570, although they coexisted for a long time with the official masses of the post-Tridentine Roman Ritual because they were strongly rooted among both the laity and religious. For example, the five masses of St. Augustine were widely accepted in Catalonia as were the extension of masses into cycles or groups of 30 (*trentenarios*,

---

[54] AHN, Inquisición, lib. 735, fols. 157–8.

[55] José Luis González Novalín, "Misas supersticiosas en la piedad popular del tiempo de la Reforma," *Miscelánea José Zunzunegui (1911–1974)*, 2 vols (Vitoria: Editorial y Librería Eset, 1975), 2:1–40.

[56] Josep M. Casas Homs, "Las gracias de la misa. Creencias populares del siglo XV," *Analecta Sacra Tarraconensia* 38 (1955): 71–8.

trentenaries), like the masses of St. Gregory or St. Amador. The trentenary of
St. Amador was deeply rooted in the Crown of Aragón, and it is well documented
in testaments from the fourteenth to the sixteenth centuries. It consisted of
30 masses that were to be said on consecutive days without the priest leaving
the church; in popular opinion it was very effective in helping people avoid
purgatory.[57] Some early modern missionaries and preachers saw themselves as
needing to suppress these superstitions:

> One person wants to have a mass to cure a disease, or to get away from a job, to
> have a happy childbirth, or to achieve a thing he desires, but that it should be
> said by a priest by the name of Juan, and not Pedro or any other name; he wants
> another mass to be given in honor of and to the glory of the Holy Trinity for
> another thing he needs, and he wants it to be celebrated by a priest who is exactly
> 31 years of age, in memory of the number three.[58]

Despite their attempts to eliminate what they saw as ridiculous superstitions, in
Catalonia such practices continued in rural areas well into the eighteenth century.

Prayers designed to eliminate sterility in women or protect women during
childbirth were particularly desired. If they were also written on tiny pieces of
paper hidden in a closed fist at the moment of childbirth, it was thought that
they increased the prayer's protective qualities. Some prayers were believed to
be especially efficient because they served to both speed up labor and protect
against enemies. For one such prayer, "Against enemies" (*Contra enemichs*), to be
effective, it was necessary to keep the paper that had seven names written on it
that were to be said daily: "amen+Saboac+emmanuell+Saboac+Sala+saboac+
dordoria+amen."[59] Protecting animals' lives was also quite important and could
make the difference between living through another winter or not. The day of
the blessing of the animals by St. Anthony in a parish or chapel dedicated to
him was a prominent festival. But the blessing had an even greater effect when
the animals were taken around the church nine times, thus drawing magical
protective circles.

Although Catalan diocesan synods condemned many of these practices,
real life and experience demonstrated their durability, especially when parish
priests themselves advised or practiced healings using these strategies. In 1660
the Holy Office prosecuted Friar Amich, a Mercedarian from Barcelona, for

---

[57]   Antoni Ma. Parramon i Doll, "El treintenari de Sant Amador," *Ilerda* 38 (1977):
75–6.

[58]   ACA, *Sermón contra la superstición*, Monacales-Universidad, leg. 130, s.n., cited by
Gelabertó, "Palabra del Predicador," 309.

[59]   Biblioteca de Catalunya, *Miscel·lània de textos devocionals*, ms. 854, fol. 183v. This is
a curious, little volume; it includes many hands and was written beginning in the fifteenth
century and ending around the mid-seventeenth century.

writing formulas to preserve health and for giving them to his parishioners so that they could put them above their doors, on their windows, or on their bed's headrest to prevent evil spirits from entering. The evil eye was commonly used to explain all kinds of negative events, from losing domestic objects to individual sickness. One of the most effective remedies against it was reciting the prayer of St. Lazarus a certain number of times while in specified postures.

* * *

Although the institutional church of the Counter Reformation strove to delineate and separate the sacred from the profane and to condemn non-authorized access to the supernatural, early modern Catalans resisted this uniformity. On many occasions, and often in consensus with parish clergy, they dynamically constructed complex forms of local Catholic religiosity, and this process continued well into the eighteenth century. The everyday practices covered here show not one, but many types of Catholicism, reflecting what Carla Russo has beautifully affirmed: local religiosity incorporated the divine into the everyday mentality of the people, humanizing God so that he would feel more familiar, proving the divine power that priests proclaimed through techniques they invented.[60] The dynamic journey between this world and the supernatural realm allowed the supernatural to permeate Catalans' everyday life in the early modern period. Their faith in an invisible world populated by angels, saints, spirits, demons, and other figures was solidly rooted in the Christianity local communities shared. That "world" was a place to go to and, at the same time, an attractive deposit of supernatural resources. Whether or not the resource was divine in origin was ultimately not the most important thing. In 1649 a peasant of the bishopric of Girona, Montserrat Riu, was accused of curing the evil eye by saying certain prayers. One witness confirmed that Riu had carried out these practices and had been seen doing so several times and, supposedly, quite effectively. The witness then hurried to add, "and I don't know ... if that is of the Devil or because God our Lord has granted her such grace." The supernatural action had cured sick people. Whether the origin of Riu's magical power was Satan or God, the witness had no idea. It didn't seem important.[61]

Translated by Maria Soukkio and
Kathryn A. Edwards

---

[60]   Carla Russo, *Società, Chiesa et vita religiosa nell'ancien régime* (Naples: Guida Napoli, 1976), clxxv.

[61]   Thorndike Collection, Mss. Coll. 49, exp. 24, fol. 22r.

Chapter 3

Lived Lutheranism and Daily Magic in Seventeenth-Century Finland[1]

Raisa Maria Toivo

At Ulvila, southwestern Finland, in 1676, a widow was accused of using magic and witchcraft for fishing. When her maid testified against her, the widow asked, "But do I not mention God, too, sometimes?" The maid admitted this, citing the widow's words: "Eyes of Salmon, eyes of Pike, Eyes of all fish, look at my nets, in the name of the Father, the Son, etc." The widow thought her words were a godly prayer and a proof against her being a witch. The court, however, interpreted them as a charm and fined her, "according to her own confessions," the considerable sum of 40 markkas.[2]

Ten years later, in another trial, the same widow was accused of using, among other things, magic on her cattle and while brewing beer: she tied bells around the necks of her cows, blew in a horn when she let them out, and burned in the beer tub the straws that had been used to cover the house floors at Christmas. She was said to have given advice on how to prevent a slash-and-burn fire from breaking free by walking around it nine times. This time she explained that the bells and horn had been a children's game and her advice for keeping the fire had only meant that it should be carefully watched. The straws had been used to pre-heat the beer tub, according to a proven custom, because having been on the floor for a time they could no longer serve any other purpose; by burning them outside the house in the tub, she could prevent fire breaking out inside. The court indeed confirmed that this was a local custom. At the end of the trial the widow was acquitted.[3]

---

[1] This work has been done as a part of a research project, "The Orthodox Lutheran Confessionalism in Seventeenth-Century Sweden and its European Context," funded by the Academy of Finland, and the Academy of Finland Center of Excellence, "Rethinking Finland: History of a Society, 1400–2000."
[2] Lower court records Bielkesamlingen vol. 27:53–5. Ulvila 11–12 Sept. 1676. Sveriges Riksarkiv (SRA).
[3] Lower Court Records, Ala-Satakunta II, KO a2:183–5 and 225–9. Ulvila 21–23 & 25 Feb. 1687 and Ulvila 11–13 July 1687. National Archives of Finland [hereafter cited as NAF].

These two cases against the same woman depict the two most often used methods of defense in magic trials in early modern Finland. The cases represent the two spheres of life where negotiation and contestation over approved or forbidden customs and conduct most often took place: faith and work. These areas are also where attitudes towards daily magic can be investigated. In the religious climate of seventeenth-century confessionalist Lutheran Sweden, to which Finland belonged, communal negotiation determined whether individual deeds were deemed as magic or a normal part of rural daily life and farming; a common custom, shared by many people, was less suspect than anything odd that only one individual did. Something that could be reasonably explained as a form of diligence or carefulness was more acceptable than something inexplicable. But attitudes were not always unanimous. Many suspected witches also explained that what they did was only an old custom or a Christian prayer, but their opponents thought it outdated "popery" or superstition. Acquittal rates and linguistic changes, among other evidence, suggest that negotiations were ongoing in early modern Finland over acceptable and unacceptable magical beliefs and practices. This chapter traces some of the key components in these negotiations.

Magic, Superstition, and Witchcraft in Early Modern Finland

Magic and superstition in early modern Finland or Sweden were concepts whose content eludes precise definition. In popular thinking they were closely connected with each other as well as with concepts like witchcraft, *maleficium*, and heresy on one hand and faith and piety on the other. The clergy and other theoretical thinkers tried to draw clear boundaries between these different definitions, but they were divided among themselves and were clearly unable to enforce their views even among the secular authorities, let alone different groups or individuals in the general population. As this situation and the existing information on witchcraft and magic in Finland suggests, then, a lot of what might be called "everyday magic" existed in seventeenth-century Finland, although how much "everyday magic" was written down in the court records varied as the seventeenth century wore on. Whereas trials for harmful witchcraft (*maleficium*, in Swedish *troldom* or *förgörning*) had been an endemic feature of court records since the Middle Ages, trials for at least partly benevolent or non-harmful magic increased considerably after the 1660s.[4] Historians have found

---

[4]   In early modern Sweden lower secular court records provide the most ample body of material to follow at least some thoughts of "ordinary people," in this case the rural farmers and their families. In Sweden, the clergy was supposed to refer all major cases of superstition or magic to the secular courts, and most of the time they seem to have done so. In comparison

that at this time authorities and elites initiated benevolent or non-harmful magic cases more often than the populace. Therefore, historians have concluded that the new type of trials represented customs that were not initially considered criminal or unlawful and may have been tolerated or even approved of before the 1660s. Despite this argument, however, it has also been shown that, although the formal complaint was frequently made by some authority, such as a church vicar or bailiff, the initiative that led to the court action—the information behind the complaint—often still lay with the neighbors of the accused. Even after the 1660s, then, neighborhood interest and disapprobation was involved in magic as well as witchcraft trials.[5]

Distinguishing between harmful and harmless magical or supernatural crime was not simple in itself. *Maleficium* was, in fact, a complex mixture of supernatural harmful actions that might or might not be linked to consorting with the devil; the early modern Finns also thought that everyone had the power to do magical harm—as well as to give magical protection. This power resided in both a person's soul and body, especially in the genitals, and it was unleashed by strong emotions like envy, hatred, love, or a sense of justice.[6] Benevolent or non-harmful magic was grouped together with superstition in general, and the early modern courts in Finland used the same Swedish terms (*vidskepelse, signerij*) for both. These cases, too, were complicated and could include practices to increase the productivity of cattle or game, to predict fortunes, to gain love, and to uncover thieves, among other activities. The practices and the practitioners varied; sometimes the magic was performed on voluntary targets and sometimes

---

to the secular courts, church court and visitation materials survive fragmentarily, and they often only consist of general admonitions for the parishioners to "give up their superstitious ways." In the secular courts, however, the parishioners were asked to describe and defend their thoughts in detail. Although these court records were the product of a system led by educated authorities, they include sometimes long and usually relatively un-coerced testimonies of the parties in each case.

    5   Linda Oja, *Varken Gud eller Natur. Synen på Magi i 1600- och 1700-talets Sverige* (Stockholm: Brutus Östlings förlag Symposion, 1999); Marko Nenonen, "Finland: Witch trials," in *Encyclopedia of Witchcraft: The Western Tradition*, ed. Richard M. Golden, 4 vols (Santa Barbara: ABC-CLIO, 2006), 2.

    6   Marko Nenonen, *Noituus, Taikuus ja noitavainot Ala-Satakunnan, Pohjois-Pohjanmaan ja Viipurin Karjalan maaseudulla 1620–1700. Historiallisia tutkimuksia 165* (Helsinki: Societas Historica Finlandiae, 1992), 39–72 and "'Envious Are All the People, Witches Watch at Every Gate': Finnish Witches and Witch Trials in the Seventeenth Century," *Scandinavian Journal of History* 18:1 (1993): 77–91; Jari Eilola: *Rajapinnoilla. Sallitun ja kielletyn määritteleminen 1600-luvun jälkipuolison noituus- ja taikuustapauksissa* (Helsinki: Bibliotheca historica 2003).

on involuntary ones. All the cases shared, however, the idea that to be defined as magic, a practice had to be superstitious and forbidden.[7]

To complicate things even more, court records do not necessarily make a clear distinction between the two categories of witchcraft and magic or superstition; in fact, early modern court records routinely used both terms simultaneously. The purpose of any deed and its benevolence or malevolence is often impossible to determine. This suggests that contemporaries did not place great importance on the distinction. On the contrary, they seem often to have applied both terms at once to the same cases regardless of the precise details. One common phrase in the sentences exemplifies this attitude: *vidskepelse och förgörning*, that is, "superstition or magic and witchcraft."[8] Just as they did not think the distinction between magic and witchcraft important, it was also difficult to differentiate between magic and superstition. In the Finnish language there are and were different terms: *taikuus*, which fairly pragmatically means magic, and *taikausko*, which can mean either superstition or a belief in magic. These differences were lost in the early modern court records, which translated both of these terms—if they ever were used in court—into the Swedish *vidskepelse*.[9] As Linda Oja states, the meaning of the term *vidskepelse* changed over time; during the seventeenth-century trials it was considered connected to witchcraft and, at least in theory, related to blasphemy and demonic relationships. In the eighteenth century it took on a meaning more like the ones in modern dictionaries: a belief or practice resulting from ignorance. Whereas in the seventeenth century this magic/superstition was thought of as an evil in itself, luring people away from God, *vidskelpse*'s evolution suggests that during the eighteenth century they were considered bad because of the effort their practitioners spent on useless activities and because they reflected the poor state of education among the people.[10]

Such diversity has been discounted in much of the history on magic and witchcraft in early modern Finland. Some historians in Finland, especially Jari Eilola, have interpreted the rise of trials especially as a form of control and suppression of the populace by the elites and authorities, that is, in the context of the growing control of the sovereign power. This perspective can be termed orthodox confessionalization.[11] Certainly the number of trials for all sorts of witchcraft, and especially for magic and superstition, rose throughout the seventeenth century; but the sharpest rise took place only after Sweden

---

[7] See, for example, Oja, *Varken Gud*, for a lengthy discussion on the characteristics of these groups, but also Nenonen, *Noituus, taikuus* and "'Envious Are All the People.'"

[8] Raisa Maria Toivo, *Witchcraft and Gender in Early Modern Society. Finland and the Wider European Experience* (Aldershot: Ashgate, 2008), 40–42.

[9] Oja, *Varken Gud*, 277.

[10] Ibid., 277, 284–92; Merriam Webster Dictionary. http://www.merriam-webster.com/dictionary/superstition (accessed 28 March 2011).

[11] Eilola, *Rajapinnoilla*.

had established (and indeed begun to lose) its place as a political or military great power. The link between the rise of witchcraft or magic and superstition prosecutions and religious changes is more problematic. Whereas the Swedish Reformation was legislatively abrupt, it was put into practice only slowly. The Diet of Västerås proclaimed in 1527 that all ministers in the country were to preach the "pure religion," but only in 1593 could the clergy and political powers of the country decide what that pure religion was. Only then did the Convention of Uppsala prepare the country's religious guidelines by adopting the Apostolic, Nicaean, and Athanasius Creed and the Augsburg Confession as well as the Church Order that had been drawn up in 1571. Moreover, while the Church Order included a liturgy in Swedish, it took more time to establish a Finnish one. Even after this, Catholic influences survived in the country during the times of Sigismund (1593–99) and Christina (1632–54), Calvinist influences were strong during the reign of Charles IX (1604–11), and at the end of the seventeenth century, Pietist criticism flowed in from German and Russian areas.[12] As this process suggests, the practice and daily experience of religion changed slowly until the first really affordable catechism was published in 1666. The last event also coincides with the rise of trials for magic and superstition. Indeed a small part of these trials was concerned with what might have been considered as remnants of Catholic or Eastern Orthodox customs or even what the contemporary clergy sometimes called "pagan," although in the seventeenth century most of their practitioners thought them fully Christian.

The slow processes of the Reformation and growing confessionalization both influenced and reflected a change in the religious ideas of various groups of people. Rather than a simple process of elites educating the populace, or even the populace starting to disapprove previously accepted customs, the religious change seems to have created a need to renegotiate a growing multitude of different views on what was approved, what could be tolerated, and what needed to be suppressed.

Prayer and Superstition

As in the examples with which this article began, one of the major defenses given by those accused of magic or witchcraft was that they had not done magic but said godly prayer. Some directly claimed that they had prayed in their own words, like the widow in the example above; more often they claimed that they prayed the Lord's Prayer or recited the Creed. Sometimes they admitted to saying the Ave Maria. There are even a handful of court records from the beginning

---

[12] Kauko Pirinen, *Suomen kirkon Historia I* (Helsinki: WSOY 1991), 334; Ulinka Rublack, *Reformation Europe* (Cambridge: Cambridge University Press, 2005), 94.

of the seventeenth century that mentions the use of rosaries or a Marian cult. In these cases the records noted the rosaries almost in passing, not as a crime worth prosecuting. Indeed, one case involved an old inheritance dispute, and the records included an inventory of the inheritance in which one article listed "a couple of strings of stones that old wives at the time used to read upon."[13] These old wives' objects and attendant practices merited no further attention, and both the authorities and some members of the populace seem to have shared this attitude. No one was interested in prosecuting or persecuting them, although the local vicar was legally supposed to inform a secular court about all his flock's crimes, something even this representative of ecclesiastical authority and pure doctrine only did intermittently.[14] The old wives' rosaries were tolerated—not liked, not approved of, but at the same time not worth further notice.

At points, however, a hint of nonconformism is discernible in these cases. Then the same practices were not merely outdated customs, but customs that the old wives must unlearn. Ten years after the case in Huittinen, a woman in the neighboring parish of Ulvila was fined three marks for "practicing mariolatria during the times of church services and therefore staying away from church."[15] In addition to the fine, the woman faced spiritual correction: she was to be sent to the cathedral chapter for confession, absolution, and some guidance. This woman already considered her practices as an alternative to those of the church; although it is quite possible that she did not consider them mutually exclusive, she was treated as if she did. Later, court cases revealed that several parishioners joined to use rosary prayers to heal a woman's eyesight or as a part of calendar festivities to ensure a good harvest.[16]

It is difficult, in many instances impossible, and for my purposes irrelevant to try to determine whether the practitioners of the rosaries "really" used them religiously or magically. One can only say they claimed to have used them religiously, and others interpreted them as having used the rosaries magically. The notion that Catholic practices like those described above should be regarded as magic and witchcraft had certainly reached Sweden by this time, if not long before. Paulinus Gothus wrote in 1630 about magic/superstition (*vidskepelse*) that among the forbidden kinds of magic was "the reading of word formulas, the names of the Trinity, Jehovah, quotations from the Bible, the Lord's Prayer, Ave Maria, and the rosary saints as well as other, non-understandable words"

---

[13] *ett paar stene båndh som käringar i den tijdh hade läst opå*: Lower Court Records Ala-Satakunta I KO a3, 304v Kokemäki, 17–20 Nov. 1634. NAF.

[14] Sven Wilskman, *Swea Rikes Ecclesiastique Werk I Alphabetisk Ordningh, Sammandragit Utur Lag oh Förordningar, privilegier och Resolutioner Samt Andra Handlingar, I* (Örebro, 1781).

[15] Lower Court Records Ala-Satakunta I KO a3, 275v. Ulvila 31 March 1634. NAF.

[16] Lower Court Records. Ala-Satakunta I KO a6, 192v Huittinen 16–18 Nov. 1646; Ala-Satakunta I KO a6, 148. Huittinen 4–5 June 1646. NAF.

and the use of the sign of the cross. Gothus grouped superstition, magic, and witchcraft together in the way that was common in learned theories of magic and witchcraft of the time.[17] A Finnish sermon writer, Bishop Ericus Erici Sorolainen from Turku, in postils from 1621–25 did not make this distinction either, but he spoke of superstition as erroneous religious practice. He dismissed Catholicism as an outward cult of deeds and rituals, where "thoughts can be far away from the numerous prayers that lips may cite." He named a number of "Papist delusions," among them the elevation of the Host, the use of salt and candles, and especially the saying of rosaries, "when they can read one 150 Ave Marias and 15 Our Fathers."[18] Later, in the second half of the seventeenth century, Bishop Gezelius, Senior, also frequently warned against what he referred to as "papist superstitions" as opposed to "established old customs" without clearly defining the difference. The former were not to be tolerated and the people were to be educated out of them or punished for using them; the latter, even though not always quite orthodox, were to be tolerated because the people were attached to them and trying to suppress them might cause more disruption than the customs themselves. Gezelius only gave a couple of examples of such superstitious or "papist" customs, and they usually related to communion. He argued that Catholic conceptions of full transubstantiation might be counteracted by using white wine at communion and commanded the clergy to watch that consecrated communion bread or wine must not land in the hands of magic practitioners. He did not feel a need to address informal prayer or candle meetings, and he mentioned the use of (consecrated) salt only in passing.[19]

Towards the 1640s and 1650s, attitudes towards rosary rituals grew ambiguous. Whereas the rituals were, for the church leaders and for those who condemned them, a part of outdated Catholic superstition, they were at the same time what some recent scholars have termed "Protestant superstition"—important forms of religious experience outside the official theology.[20] The ambiguity is what

---

[17]   Laurentius Paulinus Gothus, *Ethicae Christianae Pars Prima. Thet är catechismi förste deel. Om Gudzens Lag....* (Strängnäs: 1633), 188–94.

[18]   Ericus Erici Sorolainen, *Postilla I* (Stockholm: Christopher Reusner, 1621; facsimile Helsinki: Gummerus, 1988), 7–8, 510.

[19]   *Kircko Laki* [Church Law] 1686 XI:10; Pentti Laasonen, *Johannes Gezelius vanhempi ja suomalainen täysortodoksia* (Helsinki: Suomen Kirkkohistoriallinen Seura, 1977), 227–8. A candle meeting was a type of prayer or psalm meeting that used candles.

[20]   Susan Karant-Nunn, *The Reformation of Ritual. An Interpretation of Early Modern Germany* (London: Routledge, 1997), 185–6. Also see Marc Foster, *The Counter-Reformation in the Villages* (Ithaca: Cornell University Press, 1992); Robert W. Scribner and Trevor Johnson, *Popular Religion in Germany and Central Europe, 1400–1800* (London: Palgrave Macmillan, 1996); Helen Parish and William Naphy, eds, *Religion and Superstition in Reformation Europe* (Manchester: Manchester University Press, 2002).

attracted the attention of the educated clergy, nosy neighbors, and eventually the courts. At the end of the century, the ambiguity cleared; rosary practices were generally thought outdated and lost their appeal as religious experiences. Even though there may still have been some people who continued the practice, it was no longer deemed threatening, but useless.

The rosary rituals were only one example of Christian rites that were interpreted as magic, especially when they were used for such mundane ends as healing or protecting the health of people or cattle. Prayer was another Christian practice where authorities thought that its legitimate and illegitimate forms and relationship to magic were relatively straightforward to distinguish. Other believers made such distinctions differently. Some, like Agata Pekantytär, accused of witchcraft and magic/superstition in 1676, claimed that they had only prayed, clearly showing that they thought prayer and magic opposed. Another accused witch, Wallborg Andersdotter from Lappo, confessed to "curing toothache with salt, of helping mothers in labor with natural means and by reading the Lord's Prayer upon them, but nothing else."[21] Like Agata, Wallborg was convicted, because she had admitted to using salt and prayers. The authorities interpreted the use of consecrated salt in the context of Catholic practices and the "reading" as referring to a ritualistic practice of citing powerful words over and over, like a spell rather than a prayer.

Over the next several decades people increasingly began to define the line between prayer and magic similarly, and the defense the accused used also changed accordingly. Agata's case was among the latest where the distinction between magic and Christian rites varied widely. By the end of 1680s most of the accused had learned to claim, like Jaakko Eeronpoika Karlö and his wife in 1691, that they "used fire under the cowshed threshold when cattle are let out for the first time, and against illness, as many people in the area do, but no readings."[22] Reading had become risky, whether prayers, charms, or spells. Whereas the populace was encouraged to read catechisms and sermon books, the people were not supposed to invent unauthorized uses for the prayers they learned by heart or found in books. A similar tendency to worry about both correct and incorrect religious rituals and about mechanistic understandings of ritual can be seen in the constant concern of the Swedish church that parishioners only wanted to watch communion instead of "eating and drinking it" and that when the parishioners took part they did so in a superstitious way, by repeating the holy words of the institution after their priest.[23] Although Lutheran orthodoxy claimed adherence

---

[21]   The case was summarized in Åbo Tidningar 23 Feb. 1795 (no. 8).

[22]   Lower Court Records Ala-Satakunta II, KO a7:471–73. Ulvila 10–11 Oct. 1692. NAF.

[23]   Göran Malmstedt, *Bondetro och kyrkoro. Religiös mentalitet i stormaktstidens Sverige* (Lund: Nordic Academic Press, 2002), 134–45.

to a literal interpretation of scripture, any idea of a mechanistic use of the Word by laity roused concern.

Work, Skill, and Success

In addition to religious virtue as seen through prayer, the other major way to defend oneself against an accusation of witchcraft and superstition was to claim that what one had done was in fact a common example of diligence and skill in ordinary work, preferably an old and well-known custom. The above-mentioned Jaakko Eeronpoika Karlö and his wife provided both of the popular defenses of the late seventeenth century: they claimed that they used no readings and that they only did what many people in the area did. The specifics of their case, however, highlight why certain work methods were suspicious.

Jaakko Eeronpoika Karlö and his wife had moved into the village of Preiviikki in Ulvila in the early 1680s. Their own farm throve. In terms of social relationships, however, they had regularly upset the village by claiming bigger shares in the village's common fishing and milling than their fellow villagers were willing to give them. By 1691, the situation was further complicated because Jaakko's former maid and farmhand had married and were now in the process of clearing a new farmstead for themselves next to Jaakko. After some reciprocal meddling in each other's business, Jaakko suspected his new neighbors had practiced witchcraft against his farm. The affronts had started when Jaakko's former maid, now the mistress of the neighboring farm, had thrown the remnants of a dead calf onto the roof of Jaakko's cowshed. The following summer several of Jaakko's animals died. Jaakko also suspected that his neighbors had stolen the tail of his breeder boar and castrated one of his geese when they castrated their own.[24] The neighbors admitted to doing all three things but said they were not witchcraft: they had thrown the calf refuse on the roof in order to frighten away troublesome magpies and the rest had been mere accidents. Jaakko held to his suspicions because, as he said, his neighbors could equally well have thrown the garbage on their own roof. The rest he would have believed to be accidents if he had been informed of them immediately when they happened instead of only when further damage took place. Contrary to what is said about the reserved nature of modern Finns, early modern culture valued openness and considered publicity proof of honest intentions whether concerning economy, religion, or social agreements like engagements to marry. Secrecy was only worse if it was maintained in order to avoid the suspicions of others.

---

[24]   Lower Court Records Ala-Satakunta II, KO a6:530–31v; Lower Court Records Ala-Satakunta II, KO a7:471–73. Ulvila 10–11 Oct. 1692. NAF.

By the next court session half a year later, the matter had escalated. This time Jaakko demanded that his neighbor's wife be punished because she had claimed that Jaakko's wife was a witch. Such accusations were often voiced in the heat of an argument and were then taken back, at the latest, when the defamation suit was presented in court. Jaakko's neighbors did not do so, however, holding to their claim that Jaakko's wife, Brita, was a witch. They testified that Brita performed a strange ritual every time one of her cows gave birth: she took three pieces of bread, lit a fir torch or a smaller piece of wood, and then milked all her cows, pulling their udders three times directly onto each of the pieces of bread. She ended this ritual by feeding the pieces of bread to the cow that had given birth. This cow was always milked into a special bucket from which Brita gave the newborn calf drinks three or four times. Moreover, Brita was said to put rowan tree branches on the cowshed threshold when she let her cattle out to pasture.

During a third court session a few months later Jaakko and Brita brought their current maid to testify in their defense. She told that she knew "nothing else but that Brita had often had a rowan tree branch in the cowshed ceiling," although "she did not know if it still had roots on it or not." She also stated that Jaakko himself used fire under the cowshed threshold when cattle were let out the first time, and both of the couple lit, according to an old custom in the region, a fire under the threshold against disease—and perhaps on Easter—but that neither of them used any readings. Moreover, she said she had not been taught or told to do anything "unnecessary" in the household. The word "unnecessary" was a key concept in determining the line between magic and ordinary work: magic was unnecessary, was outside the rationally arguable—it was something that was surplus to the normal. For example, the rowan tree branch on the cowshed ceiling could be cut for decoration or to use the leaves for fodder, in which case it was perfectly natural for it to be attached to the roof. If it was not cut but had roots on, the roots were in the wrong place in the cattle shed and one needed to think why they were there. In this case as in many others, the suspicion was not spelled out in any more detail, but an analogy to other cases of magic can be made. Magic often consisted of taking small things with symbolic value from one location and placing them where one wanted more of the same: a few straws of hay from a neighbor's barn to your own to ensure that your cattle would have plenty of feed; dead things among living ones to cause harm; and more living things—like young trees with roots on—where one wanted to have prosperity and fertility.[25] Most deeds and actions were open to interpretations both as diligence and better methods of work and as something extraordinary, therefore magical. The mundane character of a lot of Finnish magic comes from this possibility of double interpretation.

---

[25]   Toivo, *Witchcraft and Gender*, 36–45, 120–28.

The case presented one party claiming that throwing calf remains on someone's roof was witchcraft and another claiming that it was not. The culprits' explanation was very rational: getting rid of magpies was probably a need that could be understood by rural farmers in the court's audience. The suspicion came from the unnecessary part of the procedure: the remains were thrown onto the roof of the neighbors, not one's own roof. Jari Eilola, following Diane Purkiss, has linked witchcraft and magic stories like these with enforcing and crossing the boundaries of a household or a person's body.[26] The narrative of presenting a deed as something out of the ordinary worked in the accusation, and claiming that it actually was quite necessary and rational in the defense may be just as noteworthy for understanding the nature of early modern Finnish magic.

Brita's cattle magic had clear ritualistic characteristics. The certain number of squirts of milk on the bread, the lighting of the torch, and the tree branches and fire under the threshold when letting the cattle out as ways of fighting cattle disease all carry a certain symbolism; in the case of the numbers it was biblical, while the fire, torch, and trees had to do with folklore on how to keep away demons. Nevertheless the defense presented the milk and bread as giving nourishing food to a cow after birth and the newborn calf and the constantly lit fire as something that allowed them to work more effectively. The most interesting explanation is the way the maid described the attempt to defend cattle against disease with fire under the threshold as an old and widely known habit in the region. Apparently the audience in the court confirmed that there was such a custom, although someone, most probably the scribe, added a note to the court records that the old local custom was "unsuitable." Unlike in the case of prayers, the courts were willing and able to enter into a discussion on the usage and rationality of individual methods of work. Also, unlike in the case of prayers, the notion that a custom was old was good when it related to work. Jaakko's case is one among many others that reflects the intersection between economic needs, communal morality, and attitudes towards magic in early modern Finland.[27]

Magic and the Rural Economy

The case between Jaakko and his neighbors dragged on for almost two years, with additional complaints of theft and eventually of fistfights between

---

[26] Eilola, *Rajapinnoilla*; Diane Purkiss, *The Witch in History. Early Modern and Twentieth-Century Representations* (London: Routledge, 1996).

[27] For further examples, see Lover Court Records Ulvila 11–12 Sept. 1676; Bielkes, vol. 27, fols. 53–4v SRA; Ulvila 11–12 July 1687; Ala-Satakunta II, KO a 2:225–29. NAF; Ulvila 21–23 and 25 Feb. 1687; Ala-Satakunta II, KO a 2:183–85. NAF; Ulvila 4–5 Nov. 1695; and Vehmaa ja Ala-Satakunta II, KO a 5:367–8. NAF. For other parts of Finland, e.g. Oulu, see 8–10 and 13–14 Jan 1674, rr13:527r–7v. NAF.

the neighbors. It was evidently just dropped without any formal sentence, although the villagers, including Jaakko, his wife, and the neighboring farm's owners, went on quarrelling over fishing rights, milling rights, and use of fields, meadows, and dead animals at least until 1705. Magic's use to resolve similar disputes and gain advantage reflected common economic assumptions and farming conditions in early modern Finland.

The Lutheran catechisms taught that God had ordered the world and given everyone his fair lot; whether it was more or less, it was what God had considered adequate and therefore should not be changed. This interpretation combined with a generally protectionist understanding of economy, where the common rhetoric was that each burgher in a town and each farmer in a village should have his necessary share of the business and the produce but no more than his share. Morally, these ideas meant that everyone was responsible for preventing anyone else from suffering or being economically disadvantaged by aggrandizing themselves inappropriately, either personally or financially. It has been argued that this moral rule guided the farming economy to such an extent that profit optimization could never be an explicit aim. This did not mean that the total amount of success would always remain the same, however; profits and produce could be increased in legitimate ways, such as with skillful and diligent work. In Lutheran teaching God himself had instituted work not only as a punishment, but also as a means of salvation: keeping people occupied made sure that they had no time to sin, and it provided a way of serving God and the community. Profit and produce, if waited for patiently, could be work's reward, but causing them through magic was stealing as opposed to the just reward of godly work.[28]

In many areas of rural Finland, the ideas of profit making and market economy first spread in small-scale dairy production.[29] Accordingly, whereas many forms of magic were in daily use in early modern Finland, cattle magic

---

[28]   Gabor Klaniczay, *The Uses of Supernatural Power. The Transformations of Popular Religion in Medieval and Early Modern Europe* (London: Polity Press, 1990), 166. For further recent discussion on how the notion of limited good could have influenced other aspects of peasants' life, see Peter Henningsen, "Peasant Society and the Perception of a Moral Economy. Redistribution and Risk Aversion in Traditional Peasant Culture," *Scandinavian Journal of History* 26:4 (2001): 271–96. Also see the development of this theme in the chapter by Johannes Dillinger in this collection.

[29]   See Jorma Wilmi, "Tuotantotekniikka ja ravinnonsaanti," in *Suomen maatalouden historia I. Perinteisen maatalouden aika Esihistoriasta 1870-luvulle*, eds Viljo Rasila et al. (Helsinki: Finnish Literature Society, 2003), 176–8. The importance of dairy production for the peasant economy has been downplayed in Finnish/Scandinavian historiography to emphasize grain production. Early modern cattle were not a particularly good breed; they were poorly fed during the winter and therefore produced very little. Consequently, the purpose of the cattle was mostly to provide manure, as is emphasized in the traditional history of Finnish agriculture. Cattle still accounted for a significant share in the early modern

was one of the most common types of magic. In this area new and successful methods of work aroused suspicion, and many people considered very similar work methods as skill when used by themselves and as magic if used by their competitors. The importance of cattle is also reflected in the number of activities where cattle magic was employed. People certainly went through much trouble to ensure that calves were born safely, were of the right sex, free from disease, safe from wild beasts, well fed, and productive. Given their importance to livelihoods and status, it is not surprising that cattle, milk, and butter are recurring themes in witchcraft and magic cases.[30]

The importance of distinguishing one's own skill from the magic used by others for success is illustrated by farmers or farmers' wives accusing their neighbors of practicing magic on the farmer's field. For example, a woman suspected her neighbor of field magic because she had seen him throwing handfuls of mud from his own fields to hers. In another, villagers claimed that one farmer should be punished for driving thistle away with magical means. Many magic cases were also discussions on what skill was and what witchcraft was. Skill in accident-prone work like construction, felling wood, and hunting could mean the difference between life and death.

Another aspect of daily life that made the rhetoric of magic versus work so frequent was that a lot of the work expected from farmers was collective. Everyone knew or at least wanted to know how others performed their work. In this environment magic was easily spotted. Moreover, one person's ability to carry out his or her duties was dependent on the way others carried out theirs. With all the mechanistic views attached to magic, people also could believe that, if it was magic, it was also superstitious, untrustworthy, possibly futile, and definitely ignorant. Again, the ambiguity of attitudes increased the interest in discussing, labeling, and prosecuting magic. Nevertheless, magic was everywhere, it was daily, and it seemed to pervade all the important spheres of life and economy.

The interdependency did not only extend to field farming. The use of the forests, some common grazing land, and part of the fisheries was also thought of as collective. The villagers would have a share in the seines, and the catch would be divided according to the taxes paid. Seines and even weirs and nets caused frequent disputes as some farmers complained that others did not allow them

---

Finnish peasant farmers' economy. The peasant household could most regularly spare butter to sell for cash, to trade for other products, or to pay taxes with.

[30] A range of cases in Lower Court Records, e.g., Vehmaa ja Ala-Satakunta I: KO a4:22–23, 26v. Ulvila 4 and 6 April 1674; Ala-Satakunta II: KO a 1:425v. Ulvila 3–5 July 1683; Ala-Satakunta II: KO a6:29–32. Ulvila 21 and 23 Feb. 1694. NAF. See also *Suomen kansan muinaisia taikoja IV* (1933); the volumes on cattle magic are more than three times larger than the three previous parts of the collection together.

to take part or that their equipment was damaged.[31] Due to population growth and increasingly dense settlement in the arable lands close to the shores, success in fishing and farming was becoming more and more crucial and competitive. The correctness of these methods became important for much more immediate and tangible reasons than any confessionalist ideology, which emphasized that the sins of any one member of the community might draw the wrath of God on all. If it was magic or witchcraft, and a farmer in your village practiced it, it was practiced on the village plot, where your crop was, too; it was practiced on the village cow herd, with your cows in, too. You were involved in it. Therefore, farmers had to ascertain not only in their own mind, but also together with the community, that all the methods used were really tolerable. The situation was tangible and the consequences practical.

In eighteenth-century Finland, it was thought that the system of open fields and common work prevented the spread of agricultural innovations. Historians have since concluded that it did not, but they have debated why: because there were few major innovations to spread, because the system still allowed some individual choice to farmers, or because farmers were actually able and willing to adopt more efficient ways of doing things communally as well as individually. The amount of effort and interest vested in the negotiations on what was magic and what was proper work would seem to support the last options—that farmers were able to form opinions and accept new things as a community, even though the role of tradition in the rhetoric suggests that novelty was certainly not an aim in itself.[32]

Conclusion

The early modern Finnish debate around magic and superstition had a strong practical component and was focused around the needs of daily life, work, and faith. In addition, it shows the limits of toleration or tolerability in early modern Finnish society. As late as the beginning of the seventeenth century, clearly Catholic practices were considered tolerable in the sense that they were not worth prosecuting, even though they were censured. Only later in the century did attitudes grow stricter, first among the authorities, then gradually among various groups of farmers and the rural population. By the end of the century, it had become safer to explain one's actions and deeds using any other

---

[31] Lower Court records e.g. Vehmaa ja Ala-Satakunta I, KO a4:19. Ulvila 21–23 Jan. 1678; Vehmaa ja Ala-Satakunta KO a5:13v. Ulvila 3–4 Feb. 1679; Vehmaa ja Ala-Satakunta II, KO a15: 195–205. Ulvila 12–14 March 1701. NAF.

[32] Eino Jutikkala, *Suomen talonpojan historia. Toinen, uudistettu ja lisätty painos* (Helsinki: Finnish Literature Society, 1958); Ilkka Nummela, "Asutus, pelto ja karja," in Rasila, *Suomen maatalouden historia*, 135.

justification than religion. Explaining that they reflected mere diligence and skill became the best option, but indeed, any rational explanation was potentially acceptable. Courts as well as neighbors seem to have entered into a genuine discussion of whether the explanations for and the methods of work used by their neighbors were really rational or not. Although the methods of work were clearly important enough to merit strict scrutiny, they and the interdependent success of the community were too important to subject all practices to tight scrutiny or invariable intolerance.

The debate also evolved around the understanding of tradition. In the religious context, "old customs" were thought of as suspect—at best outdated, at worst heretical. As far as religion was concerned, tradition implied Catholicism, sometimes even paganism, which some Protestant propagandists equated. Where work was concerned, however, the use of tradition, old customs, and widely known methods were thought of more positively. That something had been going on since one's childhood was considered as proof of its acceptability. Even though widespread customs could be termed inappropriate, the extent and duration of their acceptability were reasons to believe that an individual practicing them did at least not mean harm.

Seventeenth-century magic trials reveal that considerable effort was invested in investigating the limits of toleration. Usually instances of daily magic were too important for an immediate judgment either way. The very investigations in court reflect the need of the rural population, especially farmers, to act together as a community and incorporate individual customs and thoughts into common perceptions and practice. Defining magic, work, and prayer was a daily effort of negotiation to reach a shared understanding.

Chapter 4

The Guardian Angel: From the Natural to the Supernatural

Antoine Mazurek[1]

The *Catechism of the Council of Trent* concisely summarizes Catholic doctrine about guardian angels' activities and their relationship with humanity:

> our Heavenly Father [has] placed over each of us, in our journey towards our heavenly country, angels, guarded by whose vigilant care and assistance, we may escape the ambushes of our enemies, repel their fierce attacks, and proceed directly on our journey, secured by their guiding protection against the devious tracks into which our treacherous enemy would mislead us, and pursuing steadily the path that leads to heaven.

It added that "innumerable other important services are rendered to us by the invisible ministry of angels, the guardians of our safety and salvation."[2] Above all, then, a guardian angel was portrayed in early modern Catholicism as a guide along the path of spiritual perfection. If such an angelic role expressed, from the end of the Middle Ages, an interiorized devotion like that the *devotio moderna* promoted, the Tridentine Reform ensured its establishment through the liturgical and pastoral codification that from then on would define the Catholic confession: the pages that were devoted to the guardian angel in the *Catechism of the Council of Trent* and afterwards the liturgical office of the guardian angel included in the Roman breviary.

Such codification testifies to the great desire to discipline the practices and to control the manifestations of the supernatural in early modern Catholicism. In defining the actions of guardian angels by their extraordinary character while also underlining their discretion, the *Catechism of the Council of Trent* may seem paradoxical. But the guardian angel was above all a privileged intermediary

[1] I would like to thank Kathryn Edwards for offering me the possibility of presenting the first results of my research here.

[2] *The Catechism of the Council of Trent published by command of Pope Pius the Fifth*, trans. Jeremy Donovan (Baltimore: Lucas Brothers, 1829), 333–4; original in *Catechismus ex decreto Conc. Trident. Pii V. jussu editus* (Rome: Paulum Manutium, 1566), 555–6.

enjoying an intimate position between man and God and serving as an assistant on the path to salvation. In this capacity, it was enmeshed in the questions and tensions that mark the advent of modern Catholicism: the debate over grace, the diffusion of new spiritual techniques, and the epidemic of witchcraft and possession. Although I am unable to provide a detailed analysis here of guardian angels in each of these historical phenomena, I argue that the guardian angel represents a relationship and interaction between man and God, nature and the supernatural. Such relationships were the object of an unprecedented effort of clarification in sixteenth- and seventeenth-century Europe.[3] The particular interest that the figure of the guardian angel presents is due precisely to its presence in different forms of spiritual expression—those of theology, spirituality, piety, and the liturgy—and thus its possible ability to link them. From one discourse to another, in the shifts and reversals that one statement made of another's arguments, one can read the problems and ambiguities that, in early modern Catholicism, produced a new configuration of the relationship between the natural and the supernatural.

The Guardian Angel in a System of "Pure Nature"

Many factors allow for an explanation of the guardian angel's place in Catholic religious life between the end of the sixteenth and the beginning of the seventeenth centuries. The primary one was, without a doubt, Rome's adoption of a liturgical office appropriate for a universal church and simultaneous recognition that it must accept some local liturgical variants. This situation provided a decisive impetus towards the compilation of works about and the foundation of confraternities under the patronage of guardian angels. Debates on the liturgical status of the guardian angel from the 1580s led to the decree of 1608 that placed the feast of the guardian angel and its office into the Roman breviary. In the opinion of one of the cardinals who was a member of the Congregation of Rites, there were multiple motivations.[4] To attempt to identify them, it is necessary to remember to what extent, during the first years of the seventeenth century, Rome formed a microcosm of Catholic Reform in which certain figures simultaneously played key roles in the definition of dogma, the

---

3   See Henri de Lubac, *Surnaturel. Études historiques* (Paris: Aubier, 1946). Henri de Lubac has reconstituted this history and shown that in the end theological discourse built itself around the pairing natural/supernatural.

4   Archivio della Congregazione delle Cause dei Santi, Positiones decretorum et rescriptorum, no. 1209, fol. 1r. I have reconstructed this process in my thesis: "L'ange gardien à l'époque moderne. Culte, élaboration doctrinale et usages, XVIe–XVIIIe siècles" (PhD diss., École des Hautes Études en Sciences Sociales, 2013; forthcoming Paris: Les Belles Lettres).

regulation of religious services, and the diffusion of spirituality. Bellarmine's career path represents this interweaving. He was a member of the principal congregations of the cardinalate, in particular those of the Holy Office, Index, and Rites, and he was the spiritual director at the Roman College of the Jesuits. Along with Ludovico de Torres, Bellarmine also prepared the liturgical office of the guardian angel.

Some years ago this context led Josephine von Henneberg to propose a link between, on one hand, the foundation of a feast in honor of guardian angels and, on the other hand, the canonization process of Saint Francesca Romana which occurred that same year. In support of her argument she emphasized that the contemporary controversy surrounding grace could provide a doctrinal background for these two decisions. The controversy arose from the decree of the Council of Trent on justification, according to which human abilities alone cannot lead to justification; it requires prevenient grace through which God helps humans to transform themselves without any merit on their part. The interpretation of this decree, which deliberately did not venture a detailed explanation of this process, was the subject of numerous treatises composed principally by Spanish theologians. Dominicans and Jesuits, in particular, opposed each other so strongly that Rome decided to take control of the affair and to submit the question to the judgment of the Congregatio de Auxiliis (Congregation on Aid by Divine Grace). In 1607 they decided not to make the Tridentine decree more specific and to prohibit all publication on this subject.[5] This presumed connection between the debate over grace and the liturgical recognition of the guardian angel thus invites us to consider together the treatment that theologians gave to the figure of the guardian angel and the changes occurring in the scholastic edifice between the end of the sixteenth century and beginning of the seventeenth.[6]

During this period, theology experienced a silent revolution that reordered the way in which the relations between the natural and supernatural were considered. The system that arose, of which Henri de Lubac has magisterially shown the development,[7] asserted, on one hand, that the end of a natural being was always measured against its means, natural by definition, and on the other hand that its natural appetite could only be concerned with an equally natural good. That which relates to the supernatural, on the contrary, was always added onto nature. This new, sixteenth-century expression of the two sides of creation corresponded to a conception of divine action in the world that distinguished a

---

[5]   In particular see Paolo Broggio, *La teologia e la politica. Controversie dottrinali, Curia romana e Monarchia spagnola tra Cinque e Seicento* (Florence: Leo Olschki, 2009).

[6]   Josephine von Henneberg, "Saint Francesca Romana and Guardian Angels in Baroque Art," *Religion and the Arts* 2:4 (1998): 467–87.

[7]   Lubac, *Surnaturel* and *Augustinisme et théologie moderne* (Paris: Cerf, 2008; 1st ed., Paris: Aubier, 1965).

general concurrence, which saw God act as creator of the universe, and a specific concurrence, which saw him as intervening for individuals' salvation. Such a distinction questions the possibility of a supernatural experience and the exact function of intermediate beings in the scholastic theological model inasmuch as, since Pseudo-Dionysius, the nature of spiritual creatures is often described with the aid of terms that evoke what we mean by "the supernatural."[8]

The precise understanding of the place of the guardian angel in theological discourse necessitates reference to St. Thomas Aquinas, because most early modern Catholic writings on this topic were for the most part commentaries on Aquinas's work, and the solutions of the *Summa theologicae* would provide considerations that could become ambiguous in the theological context of the sixteenth and seventeenth centuries. In two articles in question 113 of the *Summa*, Thomas envisages angelic protection for infidels, non-baptized infants, and the future damned. For him, it is a question of defending the idea that angelic protection is given to all men because of their nature. Here is how Thomas resolves successive objections:

> Just as the foreknown, the infidels, and even the Antichrist, are not deprived of the interior help of natural reason; so neither are they deprived of that exterior help granted by God to the whole human race—namely the guardianship of the angels. And although the help they receive from there does not result in their deserving eternal life by good works, it does nevertheless conduce to their being protected from certain evils which would hurt both themselves and others ... . Angels are sent to minister, and that efficaciously indeed, for those who shall receive the inheritance of salvation, if we consider the ultimate effect of their guardianship, which is the realizing of that inheritance. But for all that, the angelic ministrations are not withdrawn for others although they are not so efficacious as to bring them to salvation: efficacious, nevertheless, they are, inasmuch as they ward off many evils.[9]

There are two parts to Aquinas's reasoning. In the first part, angelic guardianship is assigned to all of human nature and thus to all beings who share this nature, a rational nature. Natural reason is an interior aid while the guardian angel provides exterior assistance. In the other part, angelic protection is equally assigned to all humans insofar as each has blessedness as his end, and this bliss is supernatural. In the case of infidels this objective cannot *a priori* be attained, and as a consequence the role of guardian angels is limited to one of strict protection.

---

8   Lubac, *Surnaturel*, 325–428.

9   Thomas Aquinas, *Summa Theologicae*, 1st part, question 113, article 4, reply to objection 3, & article 5, reply to objection 1. English translation from Thomas Gilby, et al., trans, *Summa theologiae*, 60 vols (New York: McGraw-Hill, 1964–73).

Guardian angels are thus more shields than guides in certain circumstances. Human nature is only truly fulfilled when supernatural bliss is achieved, that is to say, when humans realize the vision of God promised to the elect or, as St. Thomas wrote in the same article, "enlightenment through doctrine" (*illuminationem doctrinae*).[10]

Faithful to the letter of the *Summa*, early modern theologians repeated that guardian angels were assigned to people to lead them to this supernatural end. But the relationship between nature and grace evolved between the time of Thomas and that of his sixteenth-century commentators. For Thomas Aquinas, the natural end of man is the vision of God (the beatific vision),[11] an end that surpasses human nature and the means at its disposal. The first masters of Salamanca at the beginning of the sixteenth century, Francisco de Vitoria and Domingo de Soto, still considered anthropology through this optic of supernatural blessedness, but with the later generation a swing occurred and the concept of the pure state of nature became a school hypothesis. Forged by scholasticism to conceive of nature independently from the gifts of grace, the pure state of nature permitted theologians to distinguish two beatitudes in humanity: a natural blessedness, by which God is known as the creator of nature, and a supernatural blessedness, that is, the vision of God. Cajetan first formulated the hypothesis in his commentary on the *Summa theologicae*. Bellarmine then used it to refute Baïus's theses by providing the foundations of a true system. But it was Suárez who would contribute the most towards its expansion by showing that its interest was above all methodological: "In order to comprehend the true condition of our nature, it is necessary to bracket off everything that goes beyond nature; for not only could it be done through intellect, but it will have been possible in itself through God; what for me is almost as certain as it is certain that all these supernatural goods are purely free."[12]

If the hypothesis is methodological, nonetheless it affects doctrine as whole, as Henri de Lubac has remarked: man is made, as a natural creature, for a blessedness that is just as natural, that is to say, proportionate to his nature. If there is thus a space for supernatural blessedness, it can only be added onto natural blessedness. For Suárez, there is no natural desire to see God. Man is not capable of an end that exceeds the limits of his nature.[13] In consequence man must be considered to be in a third state which will be added to the two others, the *status viae* and the *status patriae*: "Cajetan and the modern theologians considered that there was a third status, which they called the state of pure nature

---

[10]   Ibid., a. 5, ad. 2; see reply to objection 2.

[11]   Ibid., Ia–IIae, q. 5, a. 1.

[12]   Francisco Suárez, *De ultimo fine hominis*, disp. 15, sect. 2, §1, in *Opera omnia* (Paris: L. Vivès, 1856–78; 1st ed., Lyon, 1628), 4:146.

[13]   Lubac, *Surnaturel*, 114.

and that can be thought of as possible, although it has never in fact existed; it is even necessary to consider it in light of the intelligence of others, because this status constitutes, so to speak, its foundation."[14]

As Jean-François Courtine has written, it is "difficult to appreciate right away all the anthropological consequences that could follow from this new doctrine of a state of pure nature distinct from that of original justice."[15] The status of the guardian angel also, and especially its relationship to mankind, could only be called into question by this complete separation of the natural and the supernatural. In the *Summa theologicae*, the guardian angel was entrusted with the double task of safeguarding humanity—that is, protecting it from natural dangers and demons—and leading people into supernatural blessedness. The guardian angel thus acted on two planes, the natural and supernatural, planes that were integrated according to Thomas: preservation from natural dangers contributed in some way to achieving beatitude and angelic enlightenment preparing for the infusion of grace.[16] Even if, for St. Thomas, the angel prepared nature, including humanity, for supernatural illumination through its action, it was impossible for Aquinas to make autonomous a natural dimension of angelic guardianship.

What happened then when this coordination between the natural and supernatural was demolished? The Spanish Dominican Domingo Bañez, "founder of the modern Thomistic school,"[17] seemed to have envisaged this hypothesis first in his commentary on the *Summa theologicae*, one of the ones that had the greatest influence on seventeenth-century scholasticism. After having underlined that the text of St. Thomas did not help address this question,[18] he asserted, however, that the "methodological" work of distinguishing the domain of grace from that of nature must apply to the function of the guardian angel. Two questions then arose. The first was if the guardianship of the angels is natural or simply a benefit of grace? To answer that question it was necessary to answer another: does man in the state of pure nature need a guardian angel? These were the two questions with which one of Bañez's disciples, the Italian Dominican Giovanni Paolo Nazario (1556–1645), grappled: "Is the office of protecting men suitable to angels through nature or grace? ... If man had been put in the state of pure nature, would God have given him a guardian angel?"[19] To

---

[14] Suárez, *De gratia* (1st ed., Lyon, 1620), Prol. 4, c. 1, §2, in *Opera omnia*, 7:179, cited and translated by Courtine, *Études suaréziennes*, 53.

[15] Ibid., 56.

[16] *Summa theologicae*, I[a], q. 113, a. 5, ad. 2.

[17] Lubac, *Surnaturel*, 280.

[18] Domingo Bañez, *Scholastica commentaria in primam partem angelici doctoris D. Thomae* (Lyon: Stephanum Michaelem et socios, 1588; 1st ed., Salamanca, 1584), 542.

[19] Giovanni Paolo Nazario, *Commentariorum et scholasticarum in primam partem Summae D. Thomae Aquinatis* (Cologne: Henningii, 1621; 1st ed., Venice, 1610), 496.

these questions, that Nazario first explicitly posed, Bañez had already responded yes, distinguishing two objectives in the guardian angels' missions. The first was to lead each person to natural blessedness; the second was to accompany him to supernatural blessedness. In the state of pure nature, man would be equipped with a guardian angel who would urge him to lead an honest and virtuous life. Besides, according to Bañez, the light of reason is sufficient to know this truth;[20] reason explained why all humanity had shared a belief in the existence of personal guardian angels since antiquity, whether they were believers or unbelievers: "The philosophers who, as we have said, knew about the guardianship of angels over men, did not understand that man derived a spiritual advantage from this guardianship and it would extend to the supernatural end. But, on the contrary, according to the philosophers, all the effects of angelic guardianship have been contained within nature's limits."[21] These two theologians were thus led to define an autonomous sphere for angelic protection that, in fact, reflected the autonomous sphere given to men. This autonomy derived its meaning from the intellectual framework supporting the theory of concurrence.

One of the objectives of the system of pure nature was to highlight human autonomy regarding the Creator by offering a resolution to the question about the presence of evil in the world.[22] This topic was inseparable from the theory of concurrence. In a universe where divine intervention was consistent with either the limits of nature or the overtures of grace, the place of guardian angels indicated that human autonomy was not total, certainly not with regard to the primary cause nor with the secondary causes, that is to say especially in this case, with good or bad angels. The study of the role of guardian angels in such a configuration thus permits us to understand certain stakes in the controversy over grace that shook Catholicism throughout the seventeenth century.

The polemics were triggered by the appearance in 1588 of *A Reconciliation of Free Choice with the Gifts of Grace, Divine Foreknowledge, Providence, Predestination and Reprobation* (*Concordia liberi arbitrii cum gratiae donis, divina praescientia, providentia, praedestinatione et reprobatione*). Its author, Luis de Molina, proposed to reconcile divine omnipotence with human free will.[23] This Jesuit theologian intended to show two things: God desires to save

---

[20]   Francesco Amico was led to describe the community that would exist between guardian angels and men in political terms. See Francesco Amico, *Cursus theologicus, scholasticus et moralis* (Anvers: G. Lesteenium, 1650), 176.

[21]   Bañez, *Scholastica Commentaria*, 543.

[22]   See Heiko Oberman, *The Harvest of Medieval Theology. Gabriel Biel and Late Medieval Nominalism* (Cambridge: Harvard University Press, 1965), 48, who notes that, in this theory, God participates in all human actions, even "deformed acts."

[23]   Luis de Molina, *Concordia liberi arbitrii cum gratiae donis, divina praescientia, providentia, praedestinatione et reprobatione* (Antwerp: Joachimi Trognaesii, 1595; 1st ed., Lisbon, 1588).

all humanity and humans can contribute to their salvation through free will. To resolve this difficult equation, since the end of the fifteenth century theologians had distinguished between sufficient and efficacious grace. These theologians declared that, by according grace to everyone, God manifested his sincere desire to save all people but equally left to everyone the freedom to contribute to their salvation or not by making sufficient grace into an effective grace, which made people effectively worthy of being saved. The controversy rested on the exact status of this grace. Molina's originality consisted of presenting this distinction by reformulating the theory of divine concurrence, which he defined from now on as simultaneous concurrence. It can "appear normal that it should be mobilized in the cosmological and noetic questions and that it progressively changed its meaning."[24] In these questions angels occupied a central place. A system where the primary cause acts with the secondary cause substituted for a system where the primary cause acts in the secondary cause. Molina summarized his version of the system thus:

> Accordingly, it must be said that God *immediately*, by an immediacy of the *suppositum*, concurs with secondary causes in their operations and effects, in such a way, namely, that just as a secondary cause immediately elicits its own operation and through it produces its terminus or effect, so too God by a sort of general concurrence immediately acts with it on that same operation and through the operation or action produces its terminus or effect. It follows that God's general concurrence is not an action of God's *on* the secondary cause, as though the secondary cause acted and produced its effect after having first been moved; rather, it is an action immediately *with* the cause on its action and effect.[25]

In this configuration, influence is described as general when God acts on all the actions of the creature and as special when God grants humans influence to carry out that which they are naturally incapable of accomplishing. This made it possible to highlight the initiative of free will in the pursuit of salvation; the grace necessary to obtain it only crowned these efforts to some extent but from the outside.

In this theoretical construction the guardian angel had a singular place. On the one hand, the figure illustrated the universality of sufficient grace. Just as sufficient grace was given to all people, in the same way angelic guardianship benefitted all humanity, including infidels, heretics, and pagans. This point could be used to suggest that infidels could possibly live honestly and, thus,

---

[24]   Jacob Schmutz, "La doctrine médiévale des causes et la théologie de la nature pure," *Revue thomiste* 1:2 (2001): 217–64, here 229.

[25]   Molina, *Concordia*, 109 (Qu. 14, art 13, disp 26, trans. by Alfred J. Freddoso, http://www3.nd.edu/~afreddos/translat/molina26.htm [accessed 7 March 2013]).

prepare to receive grace. On the other hand, although this was not always indicated explicitly, it was easy to recognize in the presentation of the action of grace that which theologians said was the action of guardian angels. Now, although the parallel between the action of angels and the action of grace was not new—it was found in Pseudo-Dionysius—the concentration on the figure of the guardian angel marked a change in perspective.

In his analysis of grace, Molina was above all interested in the final stage of grace in human free will: its "relationship with our consent."[26] He provided a detailed presentation of assistance (*auxilia*) and encouragement (*auditoria*), two qualities that were supposed to promote the elevation of human actions above the natural. In the central chapter of his work, he especially insisted on the distinction between the action of assistance by grace (*auxilia gratiae*) and action by other aids which he qualified as particular assistance (*auxilia particularia*), into which he sorted the "holy angels":

> Although all these things ... are specific assistance that helps free will, both so that it maintains itself in its duty in proportion to its freedom and also so that it, with the close presence of and help from grace, cooperates with the supernatural works that look to eternal happiness ... because neither individually nor all together are they sufficient to achieve anything without more help from God. ... they do not deserve to be called assistance through grace, to speak exactly.[27]

The benefits of providence brought many opportunities for assistance through grace, an expression of God's generosity and grandeur. Taken together they formed what Molina calls "daily assistance" (*auxilio quotidiano*).[28] If everything had been accorded freely by God in the framework of the general concurrence He owes to his creations, a framework that can legitimize the end of grace, the difference between the two types of aid rested in the fact that individual assistance needed another assistance from God, a special aid (*auxilium speciale*), so that man would regulate himself according to divine will and accomplish beneficial actions.

For Molina, the guardian angel was alongside nature and the natural forces that affirm and consolidate in order that mankind perseveres and ascends to its supernatural life. If the guardian angel represented this virtuality, this opening into supernatural life, the angel was equally supposed to facilitate access to it through its intercession and its direct intervention on the human soul. Now this

---

[26]   Xavier-Marie Le Bachelet, *Prédestination et grâce efficace. Controverses dans la Compagnie de Jésus au temps d'Aquaviva (1610–1613)* (Louvain: Museum Lessianum, 1931), 1:27.

[27]   Molina, Concordia, 148.

[28]   Ibid.

last, from which the explanation was borrowed from the Thomistic-Aristotelian model, conceded to spiritual creatures, angels as well as demons, a completely natural capacity to act on bodies and, to a certain extent, on souls.

Before continuing let me stress two consequences of this theological reflection. First, despite all the efforts of theologians to distinguish as rigorously as possible the natural from the supernatural, these same theologians, in their explanation of the concrete process of making human actions supernatural, seem otherwise to concede that the path that leads to justification does not present any such discontinuity. Thus for Francisco Zumel, a strict Thomist, "By assistance through grace it is generally meant everything that guides us in a certain way, that is useful to us for the salvation of our soul and for obtaining eternal life; and more directly or immediately something that leads us or drives us towards that end. More perfectly we can call it assistance through grace."[29] On the basis of this model certain authors accord to guardian angels the ability to provoke, by means of illuminations, the conversion of infidels. Obtaining natural blessedness here substitutes to some extent for listening to the Gospels, and in this limited case the guardian angel becomes the initiator of a movement within the person to whom he has been assigned, leaving God simply with the task of achieving it.

Secondly, the liberation of secondary causes concerns the angels as much as demons, and certain theologians, notably Jesuits, insisted strongly on the permanent presence, alongside each person, of a guardian angel and a demon; the latter had the responsibility of tempting and attacking the person, and Satan placed it there because of his design to imitate God. Suárez's conviction was in this case completely representative of the dominant opinion found among his illustrious Jesuit contemporaries Gabriel Vázquez, Gregorio de Valencia, and Martin Becan: "Lucifer looks to be equal with God and declares open war against God over human salvation ... : consequently in the same way that God designs individual guardian angels for each individual person, the devil also does the same, if he can, and assigns an adversary to everyone."[30]

If the discourse on witchcraft could influence this presentation of the two spirits—the guardian angel being elsewhere considered an especially effective protection against the noxious influence of demons or bewitchment[31]—the theological system described earlier seems an essential doctrinal foundation. Now this presentation is even more remarkable particularly because we find

---

[29]   Francisco Zumel, *Variarum disputationum, tomi tres ...* (Lyon: J. Pillehotte, 1628; 1st ed., Lyon, 1609), 2.

[30]   Suárez, *De Angelis* (1st ed., Lyon, 1620), book 8, ch. 21, §30, in *Opera omnia*, 2:1097.

[31]   Among others, see Girolamo Menghi, *Compendio dell'arte essorcistica* (Bologna: Giovanni Rossi, 1576), 484. On the oppositions and inversion that structured the discourse of witchcraft and that explained, from another point of view, this symmetrical presentation, see Stuart Clark, *Thinking with Demons: The Idea of Witchcraft in Early Modern Europe* (Oxford: Oxford University Press, 1997), 31–105.

such an opinion nowhere in the work of Thomas Aquinas, although these theologians have commented extensively on that work. In the early modern era certain Catholic theologians opposed this system vigorously, not the least of them being figures such as Bañez, Estius, and Bellarmine. This opinion implied, for its partisans, the attribution of good actions to the influence of the guardian angel and the bad to that of the individual's demon.[32] Thus throughout his career Bellarmine criticized this very doctrine and the scriptural authorities most often cited in its support. He even established a significant parallel between such a configuration and the symmetrical presentation of grace and concupiscence among the Calvinists.[33] From another perspective, this legitimized the intrusive discourse on possession and the uncertainty about possession's exact nature, divine or diabolical.

The Guardian Angel: Daily Connection with the Supernatural

In theological discourse, the guardian angel thus expressed both the potential openness of all humanity to grace and the need to strengthen the forces of free will, this in a period marked by the growing complexity of theoretical attempts to reconcile these two concepts. But this series of issues was expressed in other forms: first, in the works spreading devotion throughout Catholic Europe that drew a parallel between the omnipresence of angels and the possibility of contact with the supernatural; second, in books by clerical authors who worked to elucidate the precise moment when this contact occurred by defining the exact role of the guardian angel in a hermeneutical test known as the discernment of spirits.

Theological treatises and devotional texts insisted on the omnipresence of angelic protection. The passage of the *Catechism of the Council of Trent* cited at the beginning of this article invited authors to provide example after example of existence's invasion by guardian angels. It will suffice here to cite one of the first treatises written in Italian devoted exclusively to the theme of angelic protection. Through a long and rhythmic enumeration, Andrea Vittorelli showed the presence of angels in daily life in his *Trattato della custodia che hanno i beati angeli degli uomini* (*Treatise concerning the Protection the Blessed Angels have for Mankind*):

---

[32] Opinions were divided on this point, with one side insisting more on the role of free will than the other side.

[33] See *Bibliotheca Sanctorum Patrum* (1575–79), tomi tertii censura, cited by Peter Godman, *The Saint as Censor. Robert Bellarmine between Inquisition and Index* (Leiden: Brill, 2000), 260.

> The angelic spirits influence, from the ardor of charity that they have for us, the offices of priests, prophets, preachers, masters, counselors, captains, mounted soldiers, foot soldiers, sentries, explorers; they threaten, they make us fearful, they beat us, kill us, and teach us theology, philosophy, and medicine; they are sometimes doctors, surgeons, nurses; they serve as sailors [and] guides for travels and those involved in commerce; they cooperate in human contracts ... [34]

Described here is a veritable descent into the concrete experience of human existence. The guardian angel ends by mixing itself into even the most vile earthly realities. There was elsewhere, in the description of the office of the guardian angel, a parallel often outlined between its humility and that of Christ. Carlo Ossola has suggested how this presentation of an anonymous guardian angel found an unexpected equivalent in the eucharistic theology of Port-Royal.[35] Certainly the anonymity of guardian angels was a recurring theme in theological treatises, devotional works, and sermons of the seventeenth century. The same repetition and pervasiveness was applied to the conviction that a guardian angel was present constantly near its human beneficiary. The faithful must believe in this permanent presence, and it was thus useless for them to make it appear through magical invocations. How could they summon something that was already there? The diffusion of the belief in the possibility of seeing guardian angels and of knowing their true names was attested by inquisitorial jurisprudence, which recommended caution on this subject, underlining the risks of its use in demonic magic.[36] The bull *Coeli et terrae* of 5 January 1586, in which Sixtus V condemned judicial astrology and all other forms of divination, presented guardian angels as much more effective protectors of humanity than the stars.[37] The celebrated liturgist Bartolomeo Gavanti reminded his audience that the liturgy of the Roman breviary also invited the faithful to celebrate not only one's own angel but all other angels in order to underline more emphatically the community they form.[38] These precautions testify to the control that the

---

[34]   Andrea Vittorelli, *Trattato della custodia che hanno i beati angeli degli uomini* (Venice: Valentini, 1610), reprinted in Carlo Ossola, ed., *Gli Angeli Custodi, Storia e figure dell'"amico vero"* (Turin: Einaudi, 2004), 52–3.

[35]   Carlo Ossola, "Introduzione," in Ossola, *Gli Angeli Custodi*, xxix.

[36]   Francesco Albizzi, *De inconstantia in iure admittenda, vel non. Opus in varios tractatus divisum* (Amsterdam: Ioannis Antonii Huguetan, 1683), 527. Albizzi's work is late but reflects his practice at the heart of the Holy Office during the first half of the seventeenth century.

[37]   *Magnum Bullarium Romanum, nunc denuo Illustratum ... A Pio Quarto usque ad Innocentium IX* (Lyon: Laurentii Arnaud et Petri Borde, 1673), 2:515.

[38]   Bartholomeo Gavanti, *Thesaurus sacrorum rituum seu commentaria in Rubricas Breviarii Romani* (Anvers: Plantiniana Balthasaris Moreti, 1634), 167. At the end of the eighteenth century, another Jesuit author, Pasquale De Mattei, showed this connection in a

Tridentine church wanted to have over access to the supernatural. The guardian angel was part of an arsenal of devotions which should provide a response to requests for immediate assistance from the supernatural—as magical practices did—but within frameworks defined by the clergy.[39]

Thus, to this permanent omnipresence, anonymous and unperceived, were linked some particular moments where the intervention of the guardian angel established contact with the supernatural, in particular with the administration of the sacraments. Francesco Amico, a Jesuit theologian known most for his contribution to casuistry,[40] thus described angelic intervention in the soul during confession:

> An angel proposes to counsel a man to confess his sins. He causes to be born the representation of his sins and the punishments that he merits. Thus agitated and troubled by the angel, the imagination makes the active intellect of the man act such that it produces an intelligible species of the same sins and punishments that he merits. Thanks to that species, the possible intellect chooses the simple apprehension of these same sins and punishments and, as if it were a path thus made clear, God, by knowledge and supernatural affect, immediately illuminates the man's intellect and inflames in him a will to renounce these same sins that stem from a detestable supernatural motive.[41]

The direct action of the angel on the sole faculty that is open to him, the imagination, was the triggering element of a mechanism that finds its fulfillment in God's intervention; this added a supernatural dimension but an exterior one. The angel was not thereby a simple intermediary. Like a human endowed with free will, the angel was a naturally independent agent while God seemed almost forced to give his consent.[42]

---

chapter titled "Universalità quotidiana di benefizi de'santi angeli verso di moi" in his work on guardian angels (*La divozione ai santi Angeli Custodi* [Rome: Casaletti, 1783], reprinted in Ossola, *Gli angeli custodi*, 500). The author continued by showing how they thus differed from the saints whose grace was limited.

[39] See Jean-Michel Sallmann, *Chercheurs de trésors et jeteuses de sort. La quête du surnaturel à Naples au XVIe siècle* (Paris: Aubier, 1986).

[40] Amico (1578–1651) was a professor of theology at L'Aquila, Naples, and finally Graz.

[41] Francesco Amico, *Cursus theologicus, scholasticus et moralis* (Anvers: G. Lesteenium, 1650; 1st ed., Vienne, 1630), 180. For a long-term perspective on this topic, see Spruit Leen, *Species Intelligibilis: From Perception to Knowledge*, 2 vols (Leiden: Brill, 1995).

[42] The question then arises of knowing if the guardian angel acts freely when it protects and if it itself can benefit from this action. See Juan Martínez de Ripalda, *Brevis expositio litterae Magistri Sententiarum cum quaestionibus quae circa ipsam moveri possent et authoribus qui de illis disserunt* (Salamanca: Hyacinthi Tabernier, 1635), 135–7.

On the foundation of this permanent angelic presence, individual and universal, certain theologians attempted to grant to guardian angels the faculty of inspiring, by themselves, conversions among the infidels. Questions about the salvation of the infidels arose and grew in intensity after the discovery of America. Iberian universities were notably preoccupied with knowing how infidels who led an honest life—endeavoring to reach what was described above as a natural blessedness—could receive the faith in the absence of even missionaries. Those who first looked into the question rejected a solution that proposed the presence of guardian angels, at least in normal conditions where a transmission of the faith *ex auditu* must prevail. To appeal to them in that case would be to appeal needlessly for a miracle.[43] Opinions, however, had changed among some leading Jesuit theologians half a century later. Responding explicitly to this objection, Suárez distinguished between the miraculous and the supernatural, albeit with a certain hesitation: "Whoever will not have placed an obstacle [in the way] will be illuminated or called, either externally by human intermediaries—God having arranged events to obtain this result without any miracle through another type of free providence—or by interior light through the ministry of angels, which is not at all miraculous but belongs to supernatural providence."[44]

Suárez doubtlessly intended to underline that the intervention of guardian angels was not necessarily modeled on that found in scriptures, where often they manifested visually, the miracle being meant in Suárez as the exterior manifestation of the supernatural. The case of infidels confirmed only that the supernatural plan of divine intervention was superimposed on the natural plan where the guardian angel is active. Juan Martínez de Ripalda was more explicit on this point. As in the case of the sacrament of confession described by Francesco Amico, the explicative scheme was that of divine intervention but in the final stage of an operation conducted from cover to cover by the guardian angel.[45] God's action seems uniquely to accompany the movement of secondary causes to their end, once again in accordance with the lesson of the theology of concurrence. But the nature of these interventions could be ambiguous insofar as the means used by the angels to direct the faithful soul were in every way the same as those the demons used to seduce it. It is understandable then that authors involved in spiritual direction, notably the Jesuits, attempted

---

[43]   This is the opinion of the first generation of leading theologians at Salamanca, such as Domingo de Soto or Andrés de Vega. See the solution provided by Miguel de Palacios, who was active in the second half of the sixteenth century, in his *Disputationes theologicae in secundum-tertium librum Sententiae* (Salamanca: Ildefonsi à Neyla, 1574), 148.

[44]   Francisco Suárez, *De praedestinatione et reprobatione*, book IV, ch. III, §19, in *Opera omnia*, 1:495.

[45]   See Juan Martínez de Ripalda, *De ente supernaturali disputationes in universam theologiam* (Lyon: Haered. Petri Prost, Philippi Borde et Laurentii Arnaud, 1645), 2:462–3.

to avail themselves of the figure of the guardian angel within the framework of the discernment of spirits.

The Guardian Angel and Spiritual Discernment

The guardian angel owed its diffusion in devotion and spirituality principally to the Company of Jesus. One cannot here retrace the dissemination, starting with the *Méditations sur les saints anges* of St. Louis of Gonzaga, that led to the multiplication of works on this subject notably during the seventeenth century.[46] Instead, this section will focus on a less known manifestation of the interest the Jesuits brought to the figure of the guardian angel: its use in the discernment of spirits.

As Jacques Le Brun has recalled recently, from its origins an ambiguity existed in the practice of the discernment of spirits. In the scriptural texts it indeed appeared both as a work of distinction between two exterior spirits and as discernment between good and evil where it was more the person himself, his interiority, that was the object of discernment rather than the spirits who influenced him. This tension was also reflected in the problematic place of the discernment of spirits during the early modern era as it related to the diffusion of techniques for interior meditation and the problem of diabolic possession that occurred at the same time.[47]

The use of the guardian angel can be explained both by the Jesuits' adoption of Molina's theology—to the extent to which it was coherent with their spiritual

---

[46] See Joseph de Guibert, *La spiritualité de la Compagnie de Jésus. Esquisse historique* (Rome: Institutum Historicum S.I., 1953). The role of the Jesuits in the diffusion of devotional literature focused on the guardian angel cannot be reduced simply to their role in the diffusion of devotional literature more generally. Taking into account the effect of the *Spiritual Exercises* in the Jesuit tradition and the orality of the spiritual practice that the *Exercises* presume, as well as the limits that they set on the exercants' written activity, the figure of the guardian angel must be considered as being invested by these authors with a precise function in the resolution of the problem of the status and even the legitimacy of a spiritual literature, and from this, linked quite closely to the first, of authority in the relationship between the director and the person being directed. For the methodological aspects, see Pierre-Antoine Fabre and Antonella Romano, eds, *Les jésuites dans le monde moderne. Nouvelles approches*, special edition of the *Revue de synthèse* 2–3 (1999): 455–61; Pierre-Antoine Fabre, "Sciences sociales et histoire de la spiritualité moderne: perspectives de recherche," *Recherches de science religieuse* 97:1 (2009): 33–51.

[47] See Jacques Le Brun, "Discernement des esprits, discernement spirituel, discernement intérieur," in *Le discernement spirituel au XVIIe siècle*, ed. Simon Icard (Paris: Nolin, 2011), 95–103; Moshe Sluhovsky, *Believe Not Every Spirit. Possession, Mysticism, and Discernment in Early Modern Catholicism* (Chicago: University of Chicago Press, 2007).

anthropology[48]—and by their tendency to spread the practice of discernment to everyone. Opposed to the individual experience of illuminism, the guardian angel underlined that contact with the supernatural was most often mediated. In directing the faithful, the guardian angel supported a devotion that must fill their life. The difficulty came from the very point that has arisen throughout this chapter: the guardian angel was defined as an exterior aide but it acted directly on a faculty, the imagination, which allowed access to interiority. From this fact the presentation of its role wavered permanently between two levels, one which gave it too much status to the detriment of God or another which elevated it over human free will.

A valuable example of this tension is a work that is singular on several levels, the *Tractado do Anjo da Guarda* of the Portuguese Antonio Vasconcellos, published in 1621.[49] It was a veritable summa of spirituality, doubtlessly destined for novices and entirely built around the figure of the guardian angel. In the first volume the exercise of discernment was described within the framework of the "election of the style of life," the moment that must decide the vocation of the faithful, according to the terminology of the *Spiritual Exercises*. According to a general tendency in the interpretation of the *Spiritual Exercises*, Vasconcellos seemed to reduce the exercise of discernment, which was included explicitly in the tradition of the *Exercises*, to that of prudence.[50] But he then defined it in an original way.

The book began with an enumeration of angelic perfections, notably their will and freedom, and Vasconcellos supported his argument through a plethora of scholastic and patristic texts. According to Vasconcellos, humans lacked all these perfections, especially "supernatural prudence," which the angels showed when they faced the choice of choosing between God and Satan. The assistance of the guardian angels was thus necessary throughout human existence but especially at the moment when a "style of life" was chosen. Vasconcellos showed "how the guardian angel seeks to let us move to the state most adapted to our perfection."[51] The means of its interventions were interior inspirations. The description that the author provided about them fits with the new theological system that I have described:

[48]   See Michel de Certeau, "La réforme de l'intérieur au temps d'Aquaviva, 1581–1615," in *Dictionnaire de spiritualité ascétique et mystique* (Paris: Beauchesne, 1973), vol. 8, cols. 985–1016. See the theological interpretation of the *Spiritual Exercises* by Achille Gagliardi, *S.P. Ignatii de Loyola de discretione spirituum regulae explanatae* (Naples: P. Androsii, 1851), esp. 24–8.

[49]   Antonio Vasconcellos, *Tractado do Anjo da Guarda*, vol. 1 (Evora: Francisco Simões, 1621); vol. 2 (Lisbon: Francisco Simõens, 1622). Each of the two volumes has approximately 1,000 pages.

[50]   Ibid., 1:143.

[51]   Ibid., 2:724.

To understand how the angels cause these inspirations in our soul (supposing that they cannot cause them immediately in the intellect), it is necessary to understand that they can do so in the internal sense as it wants by moving, arranging, and applying phantoms and semblances of things. Hence it arises that angels are the cause of inspirations to the extent that they make the imagination imagine something and that afterwards understanding follows it and, if it is necessary, ask that God illuminates it with the new light of grace.[52]

The guardian angel here appeared as the soul's steward and acted exclusively on its natural devices.[53] The angel "asks for" divine intervention but only "if it is necessary." This raised then the question of the discernment of these inspirations and of the role of the guardian angel in this difficult exercise. Here is how Vasconcellos resolved it:

And the angels are not content with illuminating our understanding and moving our will to bring a cure to our souls, but knowing how harmful interior movements are, that the devil causes them in us often to deceive us and make us fall off spiritual precipices without the protection of devotion and heavenly illumination, the guardian angel takes care to obtain a heavenly light for us to discern movements, and it very often manifests miraculously in assuming a body to speak to us and disabuse us [of demonic deceptions].[54]

The titulary spirit had an ambivalent function: author of inspiration in competition with the devil, it also furnished the criteria for judging inspiration's truthfulness. The first criteria was interior: the heavenly light that the guardian angel needed for the person it protected but of which it was also the intermediary dispensing this light when it was received. The second was exterior: the angel appeared "miraculously" to comfort and reassure the devout. Vasconcellos repeated it often in his work: "the interior inspiration which the angels provide manifests through apparitions and exterior warnings."[55] Doubtlessly this was the most problematic aspect of the Jesuit's work. But his presentation of an angel that totally informed the action of the person it protected was equally so: "the presence of the Angel arranges all of our actions."[56] It is uncertain if the guardian angel left a role here for free will.

---

52 Ibid., 1:736.

53 Jacques Hautin, author of another treatise on a subject similar to that written by Vasconcellos, used the metaphor of the theater explicitly. See *Angelus custos seu de mutuis Angeli Custodis, et clientis Angelici officiis tractatus* (Anvers: Ioannem Cnobbaert, 1636), 218.

54 Vasconcellos, *Tractado*, 1:701.

55 Ibid., 2:693.

56 Ibid., 2:792.

Vasconcellos's work had a limited influence. It was not reprinted, and there are few copies available in libraries. He played an important role, however, in the development of young Jesuits in the Portuguese province, and similar attempts by other authors demonstrate that his work developing these ideas was far from isolated.[57] Other indications show that the guardian angel was a familiar figure in certain spiritual circles animated by the Jesuits. But a growing suspicion marked the attitude concerning such spirituality,[58] and claims to speak with guardian angels or to have visions were condemned from 1612. Nevertheless, other evidence exists for the entire first half of the century.[59]

The confrontation of these different statements allows several points to be emphasized. First, the almost insoluble character of the theological debates about grace freed the guardian angel to some extent. In theological discourse, when confronted with the problem of the missions, the guardian angel could substitute for the figure of the missionary and bring the faith that saves to those who had never heard it. In the context of the discernment of spirits, as Vasconcellos's text illustrates, we move very easily from the angel's continuous presence near the faithful to its complete control of their actions.

Second, the description given of angelic intervention in the soul by spiritual works testified to their authors' difficulties in reconciling theological characteristics with the exigencies of experience. Thus, describing the spiritual creation in conformity to Aristotelian-Thomistic teaching as a natural being, operating naturally, raised the question of the exact role accorded to it in interiority, a place in which from now on the manifestations of the supernatural were more and more entrenched and to which a spiritual creature only had access obliquely. The recourse to miraculous apparitions is revealing, from this point

---

[57] See especially Jacques Hautin, *Angelus custos seu de mutuis Angeli Custodis*.

[58] See Michel de Certeau, *La fable mystique 1. XVIe-XVIIe siècles* (Paris: Gallimard, 1982); Sophie Houdard, *Les invasions mystiques. Spiritualités, hétérodoxies et censures au début de l'époque moderne* (Paris: Les Belles Lettres, 2008); Sluhovsky, *Believe Not Every Spirit*.

[59] See the letter of general Aquaviva dated 20 October 1612 in Archivium Romanum Societatis Iesu, Med., 23, I, fol. 272v, cited by Gianvittorio Signorotto, "Gesuiti, carismatici e beate nella Milano del primo Seicento," *Finzione e santità tra medioevo et età moderna*, ed. Gabriella Zarri (Turin: Rosenberg et Tellier, 1991), 177–201. The same author provides other examples from the first part of the seventeenth century in northern Italy. Some of these authors are Gregorio Ferrari, *Racconto di alcune cose di edificazione della vita di una serva di Dio, cui si è compiaciuta di comunicarsi straordinariamente Sua Divina Maestà* (Biblioteca braidense Milan, ms. ADXI 34 [1636–1640]) and Alberto Alberti, *Racconto delle virtuose azioni di Donna Margherita Vasques Coronado, di casa Coloma, castellana di Milano, con morali applicationi per utilità di chi legge* (Biblioteca Nazionale Centrale di Roma, Fonds jésuite, ms. 843).

of view, but as a result it provides an extraordinary manifestation of a normal modality of contact with God.

The meaning of the concept of the supernatural is particularly complex in Western thought, and Benson Saler has shown that the extension of its use into fields other than that of theology has led to some confusion.[60] The representation of the guardian angel has appeared to us as a privileged vantage point in the evolution of relations between the natural and the supernatural in early modern Europe. In explicitly basing the presence of the guardian angel in nature while assigning to it the function of opening a way to the supernatural, theologians doubtlessly brought even closer the idea of a titulary spirit for individuals, a theme that thereafter remained strong in the piety of many. It should not be surprising then that spiritual discourse, engaged in a reflection on the nature of religious experience and the practical expression of theological discourse, brought to the light of day the difficulties in such a presentation.

Translated by Kathryn A. Edwards[61]

---

[60] Benson Saler, "Supernatural as a Western Category," *Ethos* 5:1 (1977): 31–53.

[61] I would like to thank Jeff Persels and Kathleen Comerford for their help with this translation.

Chapter 5

False Sanctity and Spiritual Imposture in Seventeenth-Century French Convents

Linda Lierheimer

Sometime in 1616, a young girl arrived in the town of Roanne en Forez, a bustling town on the Loire river in central France. There she rented a room and took the name "Mademoiselle" Chaussin, although she was in fact of humble origin. The girl began to "play the *dévote*," attending the church of the local Jesuit college and taking frequent communion. After a few months she enlisted the support of the Jesuit fathers for her plan to start an Ursuline convent and said she would provide 15,000 livres towards the new foundation. Her offer was accepted, and she assembled some girls from the town to be the first nuns in the new congregation. In fact, this so-called "Mademoiselle" had been employed as a servant at the Ursuline convent in Clermont before coming to Roanne. She was unmasked when a visitor to the town recognized her and reported to her former employer, the mother superior of the convent. The servant had a history of demonic possession, and the convent had taken her in out of charity after she had been "cured" of her affliction (i.e., exorcized). The superior wrote to the girl's confessor warning him that she was a fake, and "Mademoiselle" fled the town.[1]

This chapter examines such cases of false sanctity and their relationship to convent life during the first half of the seventeenth century. Using examples from France and the Franche-Comté, a French-speaking Habsburg province on the border of France, I suggest that most cases of false sanctity were quite ordinary occurrences that, far from becoming *causes célèbres*, were addressed within convents and local communities and never reached the courts. As such, they provide insight into what Kathryn Edwards and Susie Sutch have called "the mix of the ordinary and extraordinary" in early modern European religious life.[2] My focus is not on the nature of the spiritual experiences of those accused of false sanctity, but rather on how determinations of false sanctity were negotiated within specific contexts and according to the demands of the daily life of convents

---

[1]   *Mémoires de la Mère Micolon*, ed. Henri Pourrat (Clermont-Ferrand: La Française d'Edition et d'Imprimerie, 1981), 122–5.

[2]   Kathryn A. Edwards and Susie Speakman Sutch, *Leonarde's Ghost: Popular Piety and "The Appearance of a Spirit" in 1628* (Kirksville, MO: Truman State University Press, 2008), 2.

and local communities. Distinctions between "false" and "authentic" sanctity were made in relation to material conditions, internal convent dynamics, and issues of patronage and religious politics, and mother superiors played a central role in making such distinctions.

I use the term "false sanctity" broadly to refer to anyone whose outward expressions of devotion or unusual religious gifts were deemed inauthentic. Words like "false" or "simulated" sanctity and "spiritual imposture" were used by external judges to control, define, and restrict extraordinary religious experiences, and it is important to emphasize that these categories did not necessarily describe the ways that those accused saw themselves. While there were some examples of fraud, in many cases the experiences of those accused of false sanctity were seemingly identical to those whose experiences were deemed authentic. The story of Mademoiselle Chaussin, which opens this chapter, bears a striking resemblance to that of the mother superior who condemned her, Antoinette Micolon, who herself had established the first Ursuline convent in the region by gathering a group of devout girls and gaining the support of local clergy for her project.[3] And while Chaussin's history of demonic possession might seem to disqualify her from holiness, in fact there were some, most famously Jeanne des Anges, the prioress and former *possédée* of the Ursulines of Loudun, who made the transition from demonic possession to holiness.[4] As Moshe Sluhovsky observes, "demonic and divine possession were two facets of the same religious experience," and there was no clear line between the two.[5] False sanctity was also closely connected to the new mystical trend in spirituality during this period, which, in turn, was strongly associated with women. While men could be and were accused of false sanctity, the phenomenon was profoundly gendered. Most accusations were directed at women, but perhaps more importantly, socially constructed ideas about male and female nature permeated both theologians' conceptions of false sanctity and the ways that aspiring holy women understood and pursued their relationship with God.[6]

---

[3]   *Mémoires de la Mère Micolon*, 73–83; *The Life of Antoinette Micolon*, ed. Linda Lierheimer (Milwaukee, WI: Marquette University Press, 2004), 55–65.

[4]   Michel de Certeau, *The Possession at Loudun* (Chicago: University of Chicago Press, 2000); Jeanne des Anges, *Autobiographie*, ed., Gabriel Legué and Gilles de la Tourette (Montbonnet-St. Martin, 1985).

[5]   Moshe Sluhovsky, *Believe Not Every Spirit: Possession, Mysticism, and Discernment in Early Modern Catholicism* (Chicago: University of Chicago Press, 2007), 7.

[6]   The most thorough studies of spiritual imposture and false sanctity are Anne Jacobson Schutte, *Aspiring Saints: Pretense of Holiness, Inquisition, and Gender in the Republic of Venice, 1618–1750* (Baltimore: Johns Hopkins University Press, 2001); Andrew Keitt, *Inventing the Sacred: Imposture, Inquisition, and the Boundaries of the Supernatural in Golden Age Spain* (Leiden: Brill, 2005). See also Sluhovsky, *Believe Not Every Spirit*, ch. 6. On the association of women with mysticism in this period, see Henri Brémond, *Histoire littéraire*

False sanctity was not a new phenomenon in early modern Europe, but the period did see a heightened anxiety about it, along with other types of unusual religious experience and behavior such as demonic possession and witchcraft. This anxiety was fed, on the one hand, by a new concern about unorthodox religious behavior on the part of the Counter-Reformation church, and on the other, by the religious enthusiasm and emphasis on interior spirituality that was characteristic of the age. Andrew Keitt has argued that Protestantism's rejection of miracles made the miraculous more important to Catholics as a way to distinguish themselves from Protestants. Concerns about the possibility of imposture accompanied this emphasis on the miraculous, since fraudulent saints might create skepticism about miracles in general and thus aid the Protestant cause.[7] In this context, "pretense of holiness" was defined as a form of heresy, and in many areas those accused of it were hauled before the Inquisition.[8]

In France, however, cases of false sanctity were dealt with somewhat differently than in other parts of Catholic Europe. The limited toleration granted to Protestants by the Edict of Nantes in 1598 and a long tradition in the French church of independence from Rome complicated the prosecution of heresy. There was no "French" Inquisition comparable to those in Spain, Portugal, and Venice, and ecclesiastical courts had very limited powers, which meant that false sanctity did not attract the same kind of judicial attention it did in other countries.[9] In the neighboring Franche-Comté, where the Inquisition did exist, the overlapping jurisdictions of a French-style *parlement*, seigneurial justice, and the independent rights of free imperial cities, such as Besançon, complicated its operations.[10] Other studies of false sanctity have relied heavily on Inquisition records, which tend to distort our understanding of the phenomenon by highlighting the most public and sensational cases. Judicial records, says Anne Jacobson Schutte, "are highly crafted images fashioned in accord with legal procedures, statutes, precedents, and the cultural and power dynamics of

---

du sentiment religieux en France, vol. 2 (Paris: Bloud et Gay, 1916); Barbara Diefendorf, *From Penitence to Charity: Pious Women and the Catholic Reformation in Paris* (Oxford: Oxford University Press, 2002).

[7]   Keitt, *Inventing the Sacred*, 159.

[8]   Schutte, *Aspiring Saints*, x.

[9]   In France, heresy was usually tried in secular courts. See Edward Peters, *Inquisition* (New York: The Free Press, 1988), 71, 105; Henry Charles Lea, *A History of the Inquisition in the Middle Ages* (New York: Harper, 1887), 544–5.

[10]   Lucien Febvre, *Notes et documents sur la réforme et l'Inquisition en Franche-Comté, extraits des archives du Parlement de Dôle* (Paris, 1911), 21–47; E. William Monter, *Witchcraft in France and Switzerland: The Borderlands During the Reformation* (Ithaca: Cornell University Press, 1976), 68–83. The records of the Inquisition based in Besançon have been lost: Febvre, *Notes et documents*, 22.

the past."[11] In contrast, this study uses non-legal sources, including convent histories and nuns' biographies and writings, in which false mystics and holy women appear with some frequency as minor characters who challenged the authority of the mother superior and caused disorder in the convent and surrounding area. Their stories were enacted on the mundane stage of the provincial town or convent, for the most part outside of the purview of the courts. These sources, like Inquisition records, present the phenomenon within a particular narrative frame, but they offer an alternative story in which false sanctity is presented less as a serious abomination than as a practical matter that was negotiated in the context of the day-to-day life of a community or convent. They thus complement and offer a useful corrective to the existing scholarship that relies heavily on judicial records.

The question of how to distinguish between true and false sanctity had a long history in the Catholic Church.[12] The "discernment of spirits," or the divine gift of being able to distinguish divine inspiration from diabolic illusion, was the subject of numerous theological treatises in the medieval and early modern period, the most influential of which was Jean Gerson's *De probatione spirituum* (1415). Gerson identified three possible sources for visions and other unusual religious experiences: divine, diabolic, and human. Gerson and the theologians who followed him developed what Moshe Sluhovsky has called "a new theology of the anatomy of the soul" that demarcated the "boundaries between licit and illicit forms of interiorized spirituality" and drew clear distinctions "between the divine and the demonic and between truth and falsity."[13] While these writers differed on some details, all agreed that women were particularly susceptible to false visions and revelations because of their natural moral and physical weakness.

The underlying assumption of the literature on spiritual discernment was that extraordinary religious experiences could not be taken at face value and had to be examined and assessed by the spiritual authorities. Religious writers, in turn, developed practical guidelines for how to do so. In practice, the process of determining whether such experiences were of divine origin included repeated tests and self-examination, and often involved the active collaboration of male clergy and nuns, especially if the subject lived within the walls of a convent. The emergence of a new category of "simulated sanctity" in the late sixteenth century shifted the focus of spiritual discernment from distinguishing between the

---

[11] Schutte, *Aspiring Saints*, 23.

[12] Cf. Nancy Caciola, *Discerning Spirits: Divine and Demonic Possession in the Middle Ages* (Ithaca: Cornell University Press, 2003).

[13] Sluhovsky, *Believe Not Every Spirit*, 169; see also Caciola, *Discerning Spirits*, 274–319. Caciola and Sluhovsky have recently published a synthesis of their work that allows them to bridge the medieval and early modern periods: Nancy Caciola and Moshe Sluhovsky, "Spiritual Physiologies: The Discernment of Spirits in Medieval and Early Modern Europe," *Preternature: Critical and Historical Studies on the Preternatural* 1:1 (2012): 1–48.

demonic and the divine to identifying purposeful deception and fraud.[14] In this environment, Sophie Houdard argues, the false mystic or holy woman emerged in the literature as a distinct "type," a consummate actress who performed her role for a believing public.[15]

Although Gerson warned against women who claimed the gift of spiritual discernment, in fact it was often granted to pious members of the female sex.[16] This seems to have been particularly true during the Catholic religious revival of the late sixteenth and early seventeenth centuries.[17] In a convent, distinguishing between false and authentic holiness commonly took the form of a mother superior evaluating the spiritual experiences of one or more of her nuns. An example taken from the early years of the Visitation order in France demonstrates the important role religious women, especially mother superiors, played as practitioners of spiritual discernment and in defining the boundary between true and false sanctity.[18] In November 1629, Jeanne de Chantal wrote to the mother superior of one of the convents of the new order to advise her on how to deal with novices and nuns who claimed to have extraordinary spiritual experiences: "When girls conjure up fancies regarding spiritual things, it is distressing to see the tricks and deceptions, the false visions, the imaginary raptures, the stubbornness with which they pursue austerities and other such fantasies that they claim God has given them or asked them to perform."[19] Jeanne was responding in part to a spate of group possessions of nuns that would peak with the famous case of Loudun's Ursulines in the 1630s. Her skepticism about

---

[14]   Sluhovsky, *Believe Not Every Spirit*, 189–92.

[15]   Sophie Houdard, "Des fausses saintes aux spirituelles à la mode: les signes suspects de la mystique," *XVIIe Siècle* 200 (1998): 417–32.

[16]   Dyan Elliott, "Seeing Double: Jean Gerson, the Discernment of Spirits, and Joan of Arc," *American Historical Review* 107 (2002): 26–54, here 29.

[17]   Barbara Diefendorf, "Discerning Spirits: Women and Spiritual Authority in Counter-Reformation France," in *Culture and Change: Attending to Early Modern Women*, eds Margaret Mikesell and Adele Seef (Candbury, NJ: University of Delaware Press, 2003), 242.

[18]   See Micheline Cuénin, "Fausse et vraie mystique: signes de reconnaissance, d'après la *correspondance* de Jeanne de Chantal," in *Les signes de Dieu au XVIe et XVIIe siècles* (Clermont-Ferrand: Association des publications de la faculté des lettres et sciences humaines de Clermont-Ferrand, 1993), 177–87. On the role of prioresses as practitioners of spiritual discernment, see Cynthia Cupples, "'Between You and Us': Nuns and Spiritual Discernment in Seventeenth-Century France," *Proceedings of the Western Society for French History* 33 (2005): 114–31; Diefendorf, "Discerning Spirits," 241–65 and *From Penitence to Charity*; Sluhovky, *Believe Not Every Spirit*, ch. 7; Alison Weber, "Spiritual Administration: Gender and Discernment in the Carmelite Reform," *Sixteenth Century Journal* 31:1 (2000): 123–46.

[19]   Jeanne de Chantal to une supérieure, around 24 November 1629, *Correspondance*, ed. Soeur Marie-Patricia Burns, 6 vols (Paris: Les Editions du Cerf, 1986), 3:571. For an extended discussion of Jeanne de Chantal as a practitioner of spiritual discernment, see Sluhovsky, *Believe Not Every Spirit*, 219–29.

such cases was evident in her description of these nuns as "*inventing* a thousand extravagant fantasies" and acting "*as if* they were possessed."[20]

For Jeanne, the task of the mother superior was to distinguish, not between the holy and the demonic, but between authentic and imaginary spiritual experience. In her letters, she offered guidelines to help mother superiors of Visitation convents throughout France diagnose and treat the latter. Those who entered religion young or without a true vocation were identified as at greatest risk, but all young girls were susceptible because of their lack of maturity. Superiors should watch for those who showed signs of laziness, sensuality, or extreme melancholy and keep any who claimed to have visions under close supervision. Cures included purging and bleeding to help balance the humors and preventing these girls from engaging in extreme religious practices by enforcing a strict regimen of eating, sleeping, and recreation and by limiting the amount of time they spent in prayer. In general, Jeanne advised, girls with "vivid imaginations" should not be accepted as novices.[21]

Jeanne's guidelines incorporated many of the recommendations in the religious literature regarding the discernment of spirits, though she never explicitly referenced these works.[22] However, while the latter were written by and aimed at theologians and exorcists, Jeanne's advice was intended for practical use in addressing what seemed to be a common occurrence in convent life: novices and nuns who claimed to have had mystical visions. Jeanne viewed such behavior not as a serious offense, but as an internal convent matter that could be resolved through proper guidance. To one superior who asked her advice about such cases, she replied, "As for the raptures of Sister F ... I suspect and fear them, but believe that these have a natural source and stem from the imagination ... I truly believe that this is not a case of malicious duplicity and hypocrisy."[23] In another case involving a number of girls, she urged the superior to keep the matter a

---

[20] Jeanne de Chantal to une supérieure, around 24 Nov. 1629, *Correspondance*, 3:571 (my emphasis).

[21] Jeanne de Chantal to Mère Jeanne-Charlotte de Bréchard, before 15 Sept. 1618, Ibid., 1:312; Jeanne de Chantal to Mère Anne-Catherine de Beaumont, 3 Oct. 1622, Ibid., 2:109; Jeanne de Chantal to Soeur Marie-Constance de Bressand, 25 Oct. 1625, Ibid., 2:611; Jeanne de Chantal to une supérieure, around 24 Nov. 1629, Ibid., 2:571–3; Jeanne de Chantal to Mère Marie-Aimée de Blonay, 1 April 1632, Ibid., 4:323. See also Cuénin, "Fausse et vraie mystique," 181–2.

[22] Sluhovsky argues that Chantal developed a "theology of discernment" in her letters that incorporated elements from Gerson: Sluhovsky, *Believe Not Every Spirit*, 226. It is not clear whether Jeanne had actually read Gerson's works or if the influence was indirect. In any case, the ideas of Gerson and other late medieval theologians about spiritual discernment had been absorbed into the wider religious culture by the early seventeenth century.

[23] Jeanne de Chantal to Mère Marie-Jacqueline Favre, 30 April 1616, *Correspondance*, 1:154–5.

secret, both from outsiders and within the convent walls, to keep other nuns from imitating their behavior and to avoid a scandal.[24]

Among those novices who claimed to have had visions and revelations was Marie-Constance de Bressand, who entered the Grenoble convent in 1618 at age 25. Marie-Constance came from a prominent judicial family in the region and, before joining the Visitation, had led a devout life in the world, caring for the sick and poor.[25] In her visions, God assured her of her eternal salvation and told her about the sanctity (or lack thereof) of certain people. While some, including a priest, seemed to have believed these experiences were authentic, the mother superior of the Grenoble convent, Péronne de Châtel, was doubtful and wrote to François de Sales for advice. De Sales gave his opinion that Marie-Constance was a good but misguided girl: "But as for her visions, revelations, predictions, I strongly suspect that they are useless, vain and not worthy of consideration." Based on both their frequency and content, he deemed these experiences inauthentic and drew a parallel to Nicole Tavernier, a "false prophetess" who had developed a reputation for sanctity in Paris in the late 1580s during the League, when an ultra-Catholic party took over the city.[26]

Like his close friend and collaborator, Jeanne de Chantal, de Sales saw this novice primarily as the victim of an overactive imagination. Elsewhere, he expressed concern about the effect of mystical texts on "the imagination of female readers"[27] and cited as an example a nun who "as a result of reading the books of Mother Teresa learned so well to speak like her, that she seemed to be a miniature Mother Teresa; and she believed this, imagining everything that Mother Teresa had done during her life, so that she thought she could do the same, to the point of having the same thoughts, transports of spirit, and mystical raptures just as she read that the Saint had had."[28]

---

[24] Jeanne de Chantal to Mère Anne-Catherine de Beaumont, 3 Oct. 1622, Ibid., 1:108–9.

[25] Lucien Buron, "La Mère Marie-Constance de Bressand (1593–1668)," *Revue universelle du Sacré-Coeur* 17 (Aug.–Sept. 1929), 100–110.

[26] François de Sales to la Mère de Chastel, end of 1618 or beginning of 1619, *Oeuvres de Saint François de Sales*, 27 vols (Annecy, 1892–1964), 18:324–5. On the Catholic League see Mack Holt, *The French Wars of Religion, 1562–1629* (Cambridge: Cambridge University Press, 1995), ch. 4.

[27] François de Sales to la Mère de Chastel, 13–20 June 1620, *Oeuvres de Saint François de Sales*, 19:253.

[28] François de Sales, *Entretiens spirituels*, in *Oeuvres,* ed. André Ravier (Paris: Gallimard, 1969), 1049, cited in Linda Timmermanns, *L'accès des femmes à la culture (1598–1715)* (Paris: Champion, 1993), 629. In a letter to Mère de Châtel, he similarly referred to nuns "who, after reading Mother Teresa, claim to have become as perfect and to have the same spiritual experiences as she did": François de Sales to la Mère de Chastel, 13–20 June 1620, *Oeuvres de Saint François de Sales*, 19:253.

In the early seventeenth century, publication and widespread availability of texts written by and about mystical women such as Teresa of Avila and Catherine of Siena led many readers to try to imitate their spiritual heroines. As Sarah Ferber observes, "The idea of emulating such a great figure and possibly gaining for oneself a reputation for sanctity held a potent attraction for many women."[29] Such spiritual modeling, in tandem with a new emphasis on an interiorized spirituality and its mobilization of the imagination, exemplified by the Jesuit *Spiritual Exercises*, helped spark an explosion of mystical experiences and a concurrent anxiety about the source of such experiences. As a remedy, de Sales advised the superior not to engage in discussion with Marie-Constance about whether her experiences were true or false, but to change the subject, offer examples of saints who attained sanctity through simple faith, and remind her of the importance of humility. While it was possible this was the work of the devil, he said, it was unlikely.[30]

Like many of his contemporaries, de Sales believed that "the imagination of girls [made] them more susceptible to such illusions than men."[31] Because of this weakness, women were at least partially absolved of responsibility for their actions; the worst they could be accused of was pride or self-delusion. On this basis François de Sales concluded about Nicole Tavernier, "This girl did not deceive the world on purpose, but was fooled by the devil, having no other fault but her willingness to imagine that she was holy, and using duplicity and simulation to maintain her reputation for sanctity."[32] The ambiguity of this statement (was Nicole the devil's dupe or the crafter of her own persona?) is evidence of the ambivalent reaction that such figures provoked. De Sales evoked contradictory feminine stereotypes—of artifice and duplicity, and of passivity and weakness—to defuse and contain the power and dangerous nature of these women and their activities.

In the case of Marie-Constance de Bressand, however, there was no question of demonic possession. As we have seen, de Sales dismissed the idea that the devil caused her experiences. It may be that he wished to downplay the seriousness of this case in order to preserve the order's reputation. However, he echoed Jeanne de Chantal in treating Marie-Constance's "illusions" as relatively normal

  [29] Sarah Ferber, *Demonic Possession and Exorcism in Early Modern France* (London: Routledge, 2004), 9.

  [30] François de Sales to la Mère de Chastel, end of 1618 or beginning of 1619, *Oeuvres de Saint François de Sales*, 18:326–7.

  [31] Ibid., 18:327. For more on fears about the dangers of the female imagination, see Timmermanns, *L'accès des femmes à la culture*, 622–31.

  [32] François de Sales to la Mère de Chastel, end of 1618 or beginning of 1619, *Oeuvres de Saint François de Sales*, 18:324–6. On Nicole Tavernier, see André Duval, *La vie admirable de la bienheureuse Soeur Marie de l'Incarnation* (Paris, 1621), 108–10; Brémond, *Histoire littéraire du sentiment religieux,* 2:69–71.

and advising Mère de Châtel "not to consider them strange."[33] In any case, the concerns about Marie-Constance seem to have dissipated rather quickly. In March 1620, Jeanne de Chantal wrote to Mère de Châtel, asking that she send Marie-Constance to oversee the novices at the newly established Paris convent which was struggling to survive, as it had not yet found a donor.[34] After spending five years in Paris, Marie-Constance went on to serve as mother superior of three different convents, in Moulins, Nantes, and Grenoble, before her death in 1658.[35]

Around the same time, a young nun at the Ursuline convent in Tulle, in the Limousin region, began to have visions and other extraordinary religious experiences.[36] The priest who interrogated Soeur Saint-Martial pronounced her experiences authentic, and everyone in the area considered her to be a saint. Important people consulted with her, and some claimed to have been cured by her prayers. Even monks came to receive her blessing, and the nuns in her convent adored her and confided "their most secret thoughts to her."[37] However, the mother superior of the convent, who recounted this story in her writings, was skeptical and suspected that her visions and experiences were the work of the devil.

The Ursuline convent in Tulle was founded in 1618 and, as with many new convents, its early years were difficult ones. The nuns lived in extreme poverty, and it was difficult to recruit new nuns because of rumors spread by a monk, who disapproved of the establishment, that their vows were invalid since the convent had not yet received official approval in the form of a papal bull. In the first year, the convent received only two new postulants, one of whom was Soeur Saint-Martial, the niece of a town official from nearby Limoges, whom the prioress seems to have accepted sight unseen. Two years after entering the convent, Soeur Saint-Martial began to have "many visions and revelations, speaking very learnedly about the mysteries of our holy faith and the perfections of God ... . She discussed these and other similar matters in a fashion that was admirable and surprising for a person who hardly knew how to read, or to put together two words."[38] She also claimed to have the gift of prophecy. She knew of

---

[33]   François de Sales to la Mère de Chastel, end of 1618 or beginning of 1619, *Oeuvres de Saint François de Sales*, 18:327.

[34]   Jeanne de Chantal to Mère Péronne de Châtel, 21 March 1620, *Correspondance*, 1:442. On the foundation of the Visitation convent in Paris see Marie-Ange Duvignacq-Glessgen, *L'ordre de la Visitation à Paris au XVII et XVIIIe siècles* (Paris: Cerf, 1994), 24 and Diefendorf, *From Penitence to Charity*, 174–5.

[35]   Jeanne de Chantal, *Correspondance*, 1:674; Buron, "La Mère Marie-Constance de Bressand," 105–7.

[36]   The text refers to her as a nun (*religieuse*), but it is not clear whether Soeur Saint-Martial had taken her final vows. If so, her later eviction from the convent is surprising.

[37]   *Mémoires de la Mère Micolon*, 186.

[38]   Ibid., 160, 164, 185.

the death of the Duke de Luynes during the siege of Montauban at the moment it happened,[39] saw the dismissal of the King's confessor, Father Arnoux,[40] and described in detail the death of the Bishop of Carcassonne six days before news of this event arrived in Tulle.[41]

From the beginning, the mother superior, Antoinette Micolon, had doubts about the authenticity of this nun's experiences, so much so that she kept a written record of everything that happened. She outlined the reasons for her suspicions. First, Soeur Saint-Martial seemed to have reached a state of spiritual perfection much too easily; it took saints years to attain such grace. Second, she was a bit of a show-off. Rather than trying to hide her spiritual gifts, she publicly proclaimed them. Third, there was no evidence that she had led a good life up to that point, and God normally marked those he chose for such experiences from childhood. Fourth, her prayers and devotional practices were not performed in the proper spirit. For example, she seemed much too energetic during the times that she claimed to be fasting. Finally, the mark of a true prophet or visionary was the gift of seeing the future, but this nun's predictions involved things that had already happened (the implication here is that that some kind of trickery was involved).[42] Like Jeanne de Chantal, Antoinette articulated a clear set of principles to discern false from authentic holiness, principles which, in her eyes, did not apply to Soeur Saint-Martial.

On the surface, Soeur Saint-Martial "appeared to be a perfect saint." Before long, however, she began to cause division within the convent. She disobeyed the mother superior, who had forbidden her to speak of her experiences to anyone but her confessor, by regularly recounting her visions to her fellow nuns and swearing them to secrecy. She accused another nun of having "an unruly and dangerous affection."[43] This nun believed herself to be innocent but, after speaking with her accuser, fell into despair. When the mother superior confronted Soeur Saint-Martial, she curled up into a ball under a table shouting blasphemies. She remained in this troubled state for a week, but then began to claim divine

---

[39]   Charles, Duc de Luynes (1578–1621), served as chief minister to Louis XIII and led the unsuccessful siege of Montaubon, a Protestant stronghold and center of the Huguenot rebellions, between August and November 1621. Although Micolon's account implies that he died during the siege, in fact he died of a fever in December after the siege had been lifted: Holt, *The French Wars of Religion*, 179–80; Berthold Zeller, *Le connétable de Luynes, Montauban et La Valteline* (Paris, 1879).

[40]   The Jesuit father Jean Arnoux (1576–1636) was confessor to Louis XIII from 1617 to 1621. He was dismissed on 24 Nov. 1621: Robert Birely, *The Jesuits and the Thirty Years War: Kings, Courts, and Confessors* (Cambridge: Cambridge University Press, 2003), 44–9.

[41]   *Mémoires de la Mère Micolon*, 186. Christophe de l'Estang, Bishop of Carcasonne died on 16 Aug. 1621.

[42]   *Mémoires de la Mère Micolon*, 186–7.

[43]   Ibid., 186, 187.

revelations again. Around this time, a famous Jesuit, Pierre Coton, visited Tulle and was brought to see Soeur Saint-Martial.[44] He concluded that she was deceived and her experiences were merely illusions. About six months later, the truth about the girl was finally revealed when someone who knew her informed the superior that she had led a dishonorable life before entering the convent. The nuns agreed to send her back to Limoges, her hometown, accompanied by Soeur Saint-Jean de Jérusalem, the niece of Antoinette's spiritual director, the Jesuit father de La Rénodie. Antoinette asked Soeur Saint-Jean to win the confidence of Soeur Saint-Martial by complaining about the mother superior. The disgraced sister disappeared during the journey, after confessing to her companion that she had faked her holiness: "She confessed that her fasts were not real, that since she was in charge of the cooking, she dined very well while the others were praying to God. She revealed tricks so dark and terrible that they would cause the ruin of a thousand communities."[45] Even so, many, including some of the nuns in the convent, continued to believe that a true saint had been sent away.

Like other accounts of false sanctity in religious biographies and memoirs, the story of Soeur Saint-Martial served the purpose of highlighting the virtue of the "true" holy woman in this case: the mother superior of the convent, Antoinette Micolon. Such contrasts were not mere literary tropes, but illustrated the real tensions and competition around spiritual authority within the convent. A nun who challenged the authority of the prioress by claiming superior holiness could disrupt the order of convent life, which was based on the value of obedience and allowed no room for individualism. Even a woman whose holiness was not in question could pose a problem, as Antoinette had earlier learned to her dismay when she was superior of another convent in Clermont. There she had admitted a novice who had spent a number of years practicing devotion under the guidance of the Capuchin fathers. This novice, who gained a reputation as a "divine oracle," taught the other nuns "her personal devotions and practices" and led them to reject the method of mental prayer that had been in place before her arrival.[46] The result was division within the convent, with some nuns supporting the novice and others supporting their superior. After losing this battle, largely because the novice had the support of the vicar-general of the diocese who served as the convent's male superior, Antoinette left to establish the convent in Tulle, where she encountered Soeur Saint-Martial. In the contest between Soeur Saint-Martial and Antoinette Micolon, by contrast, the mother superior emerged as the clear victor. If the reader is not convinced by the indisputable authority of

---

[44]   Pierre Coton (1564–1626) was a famous preacher who served as confessor and spiritual advisor to Henri IV and Louis XIII. See the note by Henri Pourrat in Ibid., 188.

[45]   Soeur Saint-Jean sent the superior a written account of what happened on the journey, which Antoinette later burned: Ibid., 190.

[46]   Ibid., 125.

Father Coton, the postscript to the story leaves no room for doubt: after leaving the Ursuline convent, Soeur Saint-Martial took refuge in another convent but was soon sent away. She reappeared in Tulle with an infant, whose father was a vagabond. The former nun created a scene when she came to the convent to ask for alms. "Here is your saint!" Antoinette proclaimed to the nuns, who begged their superior's forgiveness and thanked her for having delivered them from such a great danger.[47]

The case of Soeur Saint-Martial conforms to some central themes in accounts of false sanctity, especially its conventional association with sexual impurity. Absent the confession of fakery, François de Sales might have pronounced this a case of overzealous imitation, based on the nun's claim that St. Catherine appeared to her every morning. (It is unclear whether the reference is to Catherine of Siena or Catherine of Genoa, but both were important models of female sanctity in this period). The case bears some of the characteristics of demonic possession, but the narrator does not explicitly designate it as such and no exorcists were called. Rather, Mère Micolon downplays the role of the devil to portray Soeur Saint-Martial primarily as an unruly troublemaker desirous of attention. The fact that her family in Limoges sent her to a convent in another town is evidence that concerns about the girl's behavior predated her entrance to the convent and lends credence to rumors about her honor, especially considering that a relative offered 10,000 francs to the convent if they would keep her.[48] For both her mother superior and for her family the problem posed by Soeur Saint-Martial was primarily a practical matter—what to do with a girl who challenged authority and did not conform to expected behaviors—rather than a grave spiritual danger.

Another case of false sanctity, taken from the life of the Dominican nun and mystic Agnès de Jésus (1602–1634), highlights the difficulties of distinguishing divine from demonic possession. Agnès is an important figure in the religious history of the Auvergne region because of her close spiritual ties with Jean-Jacques Olier, who founded the Society of Saint-Sulpice in 1641 due in part to her influence. Soon after her death, Agnès's admirers began the canonization process that eventually led to her beatification in 1994. Thanks to these efforts, we have a number of extant accounts of her life from the seventeenth century: one by her spiritual director, Esprit Panassière; another by Arnaud Boyre, superior of the Jesuits of Le Puy, who knew Agnès well; and a third published after her death in 1665 by the Sulpician, Charles-Louis de Lantages, based on documents collected by Olier and testimonials from those who had known Agnès.[49]

---

[47]   Ibid., 191.

[48]   Ibid., 189.

[49]   Esprit Panassière, *Mémoires sur la vie d'Agnès de Langeac*, ed, B. Peyrous and J.-C. Sagne (Paris: Cerf, 1994); Arnaud Boyre, *Mémoire incomplet du Révérend Père Boyre jésuite sur la vie et les vertus de la Vénérable Mère Agnès de Jésus* (Monastère de Saint-Catherine, 2000);

Born in 1602 in Le Puy to an artisan family (her father was a knife-maker), Agnès Galand showed signs of a religious calling from an early age. As a child, she consecrated her life to the Virgin and took a vow of virginity, and she took her first communion at the unusually young age of eight. Around this time, she began to have visions, and she later received the stigmata of Jesus Christ during a religious ecstasy, though these were interior rather than visible marks.[50] In 1621, Agnès joined the Dominican Third Order, and two years later she left Le Puy for the recently founded Dominican convent of Sainte-Catherine in Langeac, where she took vows in 1625.[51] Though admitted as a lay sister, or *soeur converse*, she was quickly promoted to choir nun thanks to the lobbying of her spiritual director, Father Panassière, and other local clergy.

Around 1626, Marguerite Branche, from the nearby town of Paulhaguet, entered the convent of Sainte-Catherine in Langeac with the intention of becoming a nun. However, she left a short time later, because her parents could not come up with the required dowry.[52] Soon after leaving the convent, Marguerite began to have visions and claimed to have received stigmata on her hands and feet. People flocked from all over to see this "new saint" and kiss her wounds, "and [they] returned deceived by her, having no doubt that these were true stigmata, such as several saints have had, and that such a miracle was the mark of extraordinary holiness."[53] The convent confessor, Father Panassière, along with the other clergy in Langeac, were persuaded that Marguerite's gifts were authentic. However, Agnès was not fooled. When Panassière asked her opinion about Marguerite, Agnès scolded him for going to see her: "You have been to visit Marguerite Branche to see her stigmata, you have done wrong, and she as well by showing them to you."[54]

Charles-Louis de Lantages, *Vie de la vénérable Mère Agnès de Jésus, religieuse de l'ordre de Saint-Dominique et prieure du monastère de Sainte-Catherine-de-Sienne à Langeac* (Paris, 1863; orig. 1665). The Lantages biography has recently been reissued as *Vie de la Bienheureuse Agnès de Langeac* (Paris: Cerf, 2011). The twentieth-century editions of Panassière and Boyre are based on unpublished manuscripts preserved at the Langeac convent and the archives of Saint-Sulpice in Paris. Recently, religious scholars have begun to take an interest in Agnès. See *Mère Agnès de Langeac et son temps. Une mystique dominicaine au Grand Siècle des Ames: Actes du Colloque du Puy, 9–11 novembre 1984* (Le Puy, 1984) and *Agnès de Langeac: Le souci de la vie en ses commencements: Actes du colloque de Langeac du 15 au 17 octobre 2004* (Paris: Cerf, 2006).

[50]   Lantages places this event in Agnès's teens, citing the testimony of nuns who knew Agnès (*Vie de la vénérable Mère Agnès de Jésus*, 68). Panassière says she received the stigmata during a vision in 1626 (*Mémoires*, 259–62).

[51]   The convent was established in 1620 by nuns from the Dominican convent of Sainte-Catherine in Le Puy: Lantages, *Vie de la vénérable Mère Agnès de Jésus*, 275.

[52]   Lantages, *Vie de la vénérable Mère Agnès de Jésus*, 438.

[53]   Ibid., 439.

[54]   Boyre, *Mémoire incomplet*, 195.

As in many nuns' lives, the appearance of a "false saint" provided an opportunity for Agnès to demonstrate her gift of spiritual discernment. In this case, it was Marguerite's lack of humility, a defining characteristic of a false saint according to theologians,[55] that raised Agnès's suspicions. When Marguerite visited the convent, Agnès put her to the test: "I recognized her error the day that I met her and said to her to test her humility: 'Please, Marguerite, let me see those beautiful things' and fell to my knees before her; she willingly showed me her hands with the stigmata and said, 'Would you like to kiss these wounds?' She presented them for me to kiss; I said to myself then: 'Go, you are good for nothing'."[56] According to one biographer, Agnès took out a small knife to check how deep the wounds were. Marguerite, surprised, quickly pulled her hand away.[57]

Panassière brought in the rector of the Jesuit college of Le Puy, Father Boyre, who devised his own test intended to reveal Marguerite Branche as a fake. He gave Marguerite a document that included some serious doctrinal errors: "I induced her to commit errors against the faith of the Church ... because I did not know whether she did all this herself, so I gave her some short speeches about perfection at the beginning, and later I mixed in some meaningless points and finally some heresies, to see if the spirit that she said appeared to her would approve of things that were of no merit."[58] Marguerite returned the document to him, signed with the name "Jesus-Christ" in red letters. She told Father Boyre that God had appeared to her and approved everything in the document. When asked why he had signed in red ink, she replied that this was blood from the wound in her left hand.[59] Father Boyre then revealed the errors contained in the document. When Marguerite continued to claim that her experiences were authentic, Boyre decided to exorcize her. The exorcism was performed in the church of the convent of Sainte-Catherine and lasted for more than three weeks. Despite the best efforts of a number of priests, including Father Boyre and Father Panassière, the exorcism was unsuccessful.

The story of Marguerite Branche appears in accounts of Agnès's life largely as an indication of their subject's own spiritual gifts. Yet the similarities between the two women are striking. Both came from poor families and were unable to raise the money needed to become a nun. Like Marguerite, Agnès claimed to have received visions and stigmata and was subject to accusations of false sanctity throughout her life. When she was initially refused entry to the convent in Langeac after a wealthy benefactress changed her mind about paying Agnès's

---

55   Sluhovsky, *Believe Not Every Spirit*, 176.

56   Boyre, *Mémoire incomplet*, 195.

57   Lantages, *Vie de la vénérable Mère Agnès de Jésus*, 440.

58   Boyre, *Mémoire incomplet*, 187.

59   Lantages, *Vie de la vénérable Mère Agnès de Jésus*, 442–3.

dowry, rumors spread that she was a false *dévote*, and people shouted insults at her in the streets. They warned Agnès's parents to control their daughter, "that she was deceiving them through an appearance of devotion; and that if they were not careful, she would soon bring dishonor on them."[60] Agnès had also experienced false visions, though her biographer and spiritual director, Father Panassière, hastened to assure his readers that this could happen even to true mystics. One evening in 1621, when Agnès was at prayer, "the devil wished to deceive her" and appeared to her in a vision as a crucifix. Agnès sensed something was not right and abased herself before God, and the vision disappeared. Her confessor, Father Raboly, wondered how she could tell "good" from "bad" visions. She replied, "My Reverend Father, I know there is nothing but sin in me but I have confidence that my faithful bridegroom would not permit a poor girl who has no other desire but to love and serve him to be deceived."[61]

Raboly was not the only one who had doubts about Agnès's visions. Five years later, during the same time that Marguerite Branche was accused of having false visions, Agnès too was subjected to scrutiny. Noting the similarities between Agnès and Marguerite, the whole town, including the nuns in her convent, began to doubt Agnès's sanctity. Panassière recalled, "Everyone grumbled about Sister Agnès and me, because they were afraid that she was deceived like the other one."[62] Even Panassière began to question the authenticity of Agnès's visions. Agnès became convinced that her visions were of demonic, rather than divine, origin and, after witnessing an attempt to exorcize Marguerite, asked that she herself be exorcized. When her spiritual advisors refused, Agnès took matters into her own hands: "She took up the Ritual, and having found the page which had the adjuration against storms with several signs of the cross, she exorcized herself thinking by this means to chase away the demons."[63] Agnès's self-exorcism, used by her biographers to demonstrate her humility, was at the same time a claim of spiritual authority, an imitation of the priests whom she witnessed exorcizing Marguerite Branche. Elsewhere, Boyre recounted an incident where Agnès exorcized a servant girl in the convent who was possessed by demons.[64]

---

[60]   Ibid., 240.

[61]   Panassière, *Mémoires*, 98; see also Boyre, *Mémoire incomplet*, 38 and Lantages, *Vie de la vénérable Mère Agnès de Jésus*, 108–9. Accounts of the devil using false visions to try to trick women who had extraordinary spiritual experiences was a common trope in spiritual biographies and autobiographies of this era, as was the ability to recognize these "visions" as false. St. Teresa herself recounted how the devil attempted to fool her in this way: Teresa of Avila, *The Life of Saint Teresa of Avila by Herself*, trans. J. M. Cohen (London: Penguin, 1957), 200.

[62]   Panassière, *Mémoires*, 305.

[63]   Boyre, *Mémoire incomplet*, 38; see also Panassière, *Mémoires*, 308 and Lantages, *Vie de la vénérable Mère Agnès de Jésus*, 444–5.

[64]   Boyre, *Mémoire incomplet*, 189.

However, it is important to remember that exorcisms in the early modern period were not the monopoly of priests and were often performed by lay practitioners, many of whom were women.[65] In addition, spiritual discernment was central to the practice of exorcism;[66] thus Agnès's attempts at exorcism were a logical extension of her gift of spiritual discernment. Still, these incidents may well have been deemed suspect in a period when the church was beginning to crack down on lay exorcists.

Experts, including Father Boyre, who had "unmasked" Marguerite through trickery, pronounced Agnès's visions authentic.[67] The celebrated Jesuit priest, Father Jacquinot, while visiting Langeac likewise gave Agnès his stamp of approval.[68] But even after powerful clergy had affirmed her sanctity as authentic, rumors continued to spread that Agnès was faking. After Agnès was elected superior in 1627, a nun in the convent accused Agnès of eating secretly while claiming to live on nothing but communion wafers, asserting that "the mother had food brought secretly to her room and elsewhere, that she swallowed eggs on the sly, and that she fed herself at night with pieces of roast meat that had been placed under her pillow." Convinced of the truth of this accusation, the other nuns regarded Agnès as "a girl under the illusion of Satan" and wrote to the bishop asking that she be removed as superior.[69] The bishop came to her defense but eventually complied, and Agnès was deposed in 1630 after serving three years as superior. Eventually the nuns recanted, and Agnès was reelected superior in 1634.[70]

So what explains the fact that Agnès's experiences were deemed authentic, while Marguerite was branded a fake (*simulatrice*)? There are no real clues in the accounts by Agnès's biographers, who were clearly biased in favor of Agnès. The support of local clergy and Agnès's monastic status seem to have been deciding factors; however, it is impossible to know for sure since we have so little information about Marguerite. But clerical pronouncements in such cases did not always correspond with popular opinion. After more than three weeks of exorcism had no effect on her, some people continued to think Marguerite Branche was holy, and doubts about the authenticity of Agnès's holiness continued throughout her lifetime, evidence of how subjective the distinction between false and authentic sanctity could be.[71]

A final example of a nun who gained a reputation for sanctity in the Franche-Comté shows how economic and material circumstances could be important

---

[65]　Sluhovsky, *Believe Not Every Spirit*, 39–42.

[66]　Caciola, *Discerning Spirits*, 244.

[67]　Panassière, *Mémoires*, 271; Lantages, *Vie de la vénérable Mère Agnès de Jésus*, 447.

[68]　Panassière, *Mémoires*, 328–9; Lantages, *Vie de la vénérable Mère Agnès de Jésus*, 460.

[69]　Lantages, *Vie de la vénérable Mère Agnès de Jésus*, 536, 537–8.

[70]　Ibid., 537–42.

[71]　Panassière, *Mémoires*, 307.

factors in cases of false sanctity. The nun, whose convent had been pillaged during the Ten Years War between France and Spain, was said to have posed as a saint in order to procure alms to help repair the losses her convent suffered. The woman was celebrated as a visionary, prophet, and stigmatic, but suspicions were raised when the nun made predictions about the archbishop of Besançon, who decided to look into the matter and had her brought to Besançon. (Unfortunately, the sources do not specify what these predictions were.) When she arrived, the townspeople received her like "an angel from heaven" and showered her with gifts. The nun convinced the townspeople that she had received divine gifts, including the stigmata, and that Jesus often appeared to her.[72]

During the Ten Years War, which lasted from 1635–44, the Franche-Comté was devastated by war, plague, floods, and famine, and the French and Swedish armies ravaged the countryside.[73] One contemporary chronicler described the year 1639 as "the most fatal and tragic" in the history of the region. In Besançon, starving people resorted to cannibalism: "Wheat was scarce and very expensive, which led to many deaths; the poor slept in the streets, crying and shouting: I am dying of hunger; every morning more dead were found, sometimes as many as 30 …. Horsemeat sold at a high price, some killed men and ate them."[74] These dire circumstances inspired many popular religious displays intended to convince God to end their misfortunes. During the summer of 1641, there were almost daily religious processions by the clergy and inhabitants of Besançon, who scourged themselves in the churches and prayed for peace.[75] In such terrible times, it is not surprising that people would turn to a holy woman for guidance

---

[72]   *Chroniques de l'ordre des Carmélites en France*, 3 vols (Troyes, 1846), 3:106–7. The story is based on an account of the life of Mère Thérèse de Jésus (1581–1647), the mother superior of the Carmelites of Besançon, written by one of her fellow nuns, Catherine de l'Annonciation.

[73]   The Franche-Comté was under Spanish Hapsburg rule from 1493–1678. The Ten Years' War (1635–44) was part of the Thirty Years' War between France and Spain. In May 1636, the French army invaded the Franche-Comté and laid siege to Dole. The siege was unsuccessful, but for the next three years the province was devastated by invading armies who sacked villages and burned most of the towns in the region. The war worsened a plague epidemic that lasted from 1635–40 and sparked a subsistence crisis in the years 1637–39: Darryl Dee, *Expansion and Crisis in Louis XIV's France: Franche-Comté and Absolute Monarchy, 1674–1715* (Rochester: University of Rochester Press, 2009), 24–5. See also Jean Boichard, ed., *Le Jura* (Toulouse: Privat, 1986), 93–7 and Roland Fiétier, ed., *Histoire de la Franche-Comté* (Toulouse: Privat, 1977), ch. 8.

[74]   Jehan Girardot de Noseroy, *Histoire de dix ans de la Franche-Comté de Bourgogne (1632–1642)* (Besançon, 1843), 224–6. Girardot de Noseroy (1580–1651) was a magistrate at the Parlement of Dole.

[75]   "Etat de ce qui s'est passé à Besançon depuis 1612," in *Mémoires et documents inédits pour servir à l'histoire de la Franche-Comté publiés par l'académie de Besançon*, vol. 9 (Besançon, 1900), 226, 229.

and consolation, or that a convent might hope to profit from her reputation for holiness.

The new bishop, Claude d'Achey (1637–54), found himself in a difficult position. D'Achey was appointed archbishop of Besançon in 1637 in the midst of the Ten Years War. His main concerns were to see to the religious needs of his flock in a time of crisis and to maintain ecclesiastical discipline in the face of abuses resulting from war, disease, and famine.[76] In general, the bishop was skeptical of unusual spiritual claims: at around this time he chastised some women who claimed to have had revelations that no one should work on Saturday afternoons, and he forbade the publication of any new miracles without his permission.[77] He decided to consult the prioress of the Carmelite convent in Besançon for help in determining whether the nun's visions "came from god or the devil."[78]

The Carmelite convent in Besançon had been founded fairly recently, in 1616.[79] The first nuns had been sent from the convent in the nearby town of Dole, which in turn had been founded by nuns from Dijon, across the border in French territory. The nuns met with resistance to their plans from the city fathers, who delayed giving them permission to build their convent, which was not completed until 1622.[80] The municipal authorities were concerned about the multiplication of religious houses in the city and worried that the new convent might be a drain on the town's resources.[81] The Carmelites' success can

---

[76] M. Richard, *Histoire des diocèses de Besançon et de Sainte-Claude*, vol. 2 (Besançon, 1851), 320–22; Maurice Rey, *Les Diocèses de Besançon et de Saint-Claude* (Paris: Editions Beauchesne, 1977), 109.

[77] Richard, *Histoire des diocèses*, 323–4.

[78] *Chroniques de l'ordre des Carmélites*, 3:106.

[79] Archives départementales du Doubs, 124 H1: "Acte en bonne forme des Srs Gouverneurs de Besançon portant reception et permission d'erection du nouveau monastere des religieuses Carmelites en ladite ville sous les conditions y enoncées."

[80] "Recueil des choses les plus remarquables qui ce sont passés en la Fondation du Monastere de l'immaculée conception de la tres saincte Vierge et de sainct Joseph des Carmelites de Besançon," Bibliothèque diocésaine de Grammont, fonds carmélites, côte 2445.

[81] Marie de l'Enfant Jésus, *Carmélites d'hier et d'aujourd'hui* (Colmar-Paris: Alsatia, 1967), 52. Opposition to the Carmelites may have been due to competition with the Ursulines, who requested permission to establish a congregation in Besançon that same year. The Ursulines had to delay their foundation until 1619: J. de Trévilliers, *Sequania Monastica: Dictionnaire des Abbayes, Prieurés, Collèges et hôpitaux conventuels, ermitages de Franche-Comté et du diocèse de Besançon antérieurs à 1790* (Vesoul: chez l'auteur, 1951), 1:66–7. A few years later, the town opposed the archbishop's efforts to establish a convent of Tiercelines on the grounds that they had recently given approval for two new convents, the Carmelites and Ursulines. Abbé L. Loye, *Histoire de l'Eglise de Besançon*, 3 vols (Besançon, 1902), 3:336. Such opposition to the establishment of a new convent was not unusual in this period.

be attributed to the support of the powerful noblewoman, Caroline d'Autriche, Comtesse de Contecroix, who laid the first stone of the convent, and of their foundress, Marguerite Bareur, from a prominent local family.[82]

The Carmelite prioress, Thérèse de Jésus, had played an important role in the early years of the order in France. Born Nicole Mercier in 1581 to a noble family in Troyes, a town in north-central France, she went to Paris to seek out Madame Acarie when she heard of her plans to establish the reformed Carmelites in France. As a novice, she accompanied the Spanish mothers to establish a new convent in Dijon in 1605, moved on to found another in Dole, and finally came to Besançon, where she served as prioress for many years, including the period from 1635 until her death in 1646.[83]

At the bishop's request, Mère Thérèse met with the visionary nun in the convent parlor and was not impressed. She informed d'Achey that in her opinion the girl's behavior was either fakery or illusion. The nun promptly left town, accompanied by some of her admirers whom she rewarded by having a mystical ecstasy in their presence before continuing on her route "laden with presents." But on her way home, she was stopped by soldiers, who stole all her riches, and she returned to her convent just as poor as she had left it. A Jesuit priest was dispatched to her convent and successfully convinced the girl of the error of her ways, whereupon she made a public confession before her fellow nuns. Soon after this, her convent was forced to close down, and the nuns were dispersed to other religious communities.[84]

While the documents are silent about the identity of the nun, we know that she came from a convent in the surrounding area that the war had disrupted. The nun's superior wrote numerous letters to Thérèse de Jésus in Besançon asking for her approval of this "saint," an indication of the desperate situation the convent faced and their need for the resources the nun brought to the community. In the wake of invading armies, towns were abandoned and many parishes were left without priests. Churches and monastic houses were burned, and the monks and nuns forced to take refuge in nearby towns.[85] One such house was the Benedictine abbey of Baumes-les-Dames. This convent, which until its reform in the late sixteenth century had had a reputation for laxity, was pillaged by the Swedes in July 1637. In 1638 the nuns fled to Besançon; the abbess, Hélène de Rye, stayed behind but soon joined them and remained in Besançon until her

[82] "Recueil des chose les plus remarquables"; Philippe Maréchal, *Une cause célèbre au XVIIe siècle: Béatrix de Cusance, Caroline d'Autriche, Charles IV de Lorraine* (Paris: Champion, 1910), 178. Caroline d'Autriche, Countess of Contecroix, was the illegitimate daughter of Emperor Rudolphe II of Germany (see Maréchal, *Une cause célèbre*, 167–78).

[83] *Chroniques de l'ordre des Carmélites*, 3:57–87; Marie de l'Enfant Jésus, *Carmélites d'hier*, 55.

[84] *Chroniques de l'ordre des Carmélites*, 3:107.

[85] Richard, *Histoire des diocèses*, 310–11; Rey, *Les Diocèses de Besançon*, 117.

death in 1647. If the visionary nun was indeed from this convent, that would place the episode in the first year of d'Achey's episcopacy. The bishop would have had to handle the matter delicately, since the abbess was from one of the most prominent families in the region and the niece of his predecessor, Archbishop Ferdinand de Rye.[86]

The Carmelite convent in Besançon was experiencing great hardship as well, and scarce resources may at least partially explain why the prioress was suspicious of the visiting holy woman. If the people of Besançon gave all of their gifts to this nun, what would be left for *her* convent? At stake may also have been the Carmelites' own status as a source of the miraculous. The Carmelites possessed relics of Ana de Jesus, one of St. Teresa's companions and a founding mother of the order in France, which were known to bring about miraculous cures. When the convent's surgeon was stricken with the plague in 1636, Thérèse de Jésus sent these relics "which had already performed many miracles" to him and he was cured the next day, much to the astonishment of his doctors who had given him up for dead.[87]

The ending of the story leaves some important issues unresolved. Was this simply a case of fakery for financial gain, a kind of spiritual "con job" to use a modern term? A more likely scenario is that the mother superior of a religious community in desperate straits saw a solution to the convent's troubles in a girl with unusual spiritual gifts. The nun's willingness to submit to the Jesuit priest would seem to be an indication that she was not purposely trying to deceive her audience and has parallels with stories of other nuns whose extraordinary spiritual experiences were deemed authentic. In another context, her humility and willingness to accept that her experiences were illusions might have been offered as evidence of sanctity, rather than error. But while theoretically humility was an important factor in determining holiness, in reality politics and patronage played a critical role. Consider, for example, a similar case of a Carmelite nun in Bordeaux whose visions, prophecies, and devotional practices were deemed suspect by a priest who accused her of "taking dreams for reality" and persuaded the prioress of her convent that she was "deceived by the devil." Her relatives appealed to the faculties of theology of Bordeaux and Toulouse who gave their approval to the girl's experiences and spiritual practices. The girl regained her reputation for sanctity, and her spiritual gifts became even more profuse than

[86]   Abbé L. Besson, *Mémoire histoirique sur l'abbaye de Baume-les-Dames* (Besançon, 1845), 81–3; see also Richard, *Histoire des diocèses*, 332–34. The abbey was abandoned in 1638, and the nuns did not return until the late 1640s. On Hélène de Rye's relationship to Ferdinand de Rye, see François de Sales, *Oeuvres de Saint François de Sales*, 13:111. On the episcopacy of Ferdinand de Rye (1586–1636), see Rey, *Les Diocèses de Besançon*, 107–10.

[87]   Bibliothèque municipale de Besançon, ms. Chifflet 3, fol. 101. The surgeon, Toussaint Joliot, personally attested to this miracle.

before.[88] Without the influence of powerful relatives who lobbied on her behalf, her fate might have been no different from the nun in the Franche-Comté.

Although the literature on spiritual discernment offered guidelines for distinguishing between authentic and false sanctity, in practice such distinctions involved pragmatic as well as spiritual considerations and were informed by material conditions, internal convent dynamics, and issues of patronage and religious politics. The examples in this chapter suggest that cases of false sanctity, and, I would argue, all forms of unusual religious behavior, must be understood within their specific local contexts. In religious histories and narratives of nuns' lives, false sanctity emerges as part of the texture of convent life, a commonplace, almost mundane, occurrence, as likely to be attributed to an overactive imagination as to demonic possession. Noteworthy is a striking absence of a serious sense of spiritual danger. Depending on the circumstances, the penalties ran the gamut from gentle guidance, to public confession, to expulsion from a convent, to exorcism—but none of the cases discussed in this chapter resulted in serious consequences for the accused on the level of being hauled before the Inquisition or even a local court. Rather, they were handled at the local level by mother superiors and local clergy for whom the pragmatic and the spiritual were deeply entwined and whose chief concern was to keep order and ensure spiritual conformity within the convent. While these cases may seem a far cry from the religious *causes célèbres* of the early seventeenth century, such as the mass possession of nuns in Loudun, the latter were merely the most extreme expressions of a spiritual environment in which distinctions between the holy and the profane, the divine and the demonic, were made every day.

---

[88]   *Chroniques de l'ordre des Carmélites*, 3:266–8.

Chapter 6

Magic, Dreams, and Money

Jared Poley

In one of the more haunting definitions of a dream to be published in the early modern period, Philip Goodwin explains that "Dreames go much in the dark, as they usually be in the dark *night*, so of a darke *nature*: so vailed and covered, as they commonly require an Interpreter. A Dream is a close covered *Dish* brought in by *night* for the *Soul* to feed on; And is it not meet for a man, after to uncover the *Dish*, to see and know upon what Meat he hath eaten?"[1] Dreams provided a magical space in which supernatural knowledge could be transmitted and made useful. Dreams revealed important knowledge, but they required skilled interpretation. The subject of this chapter is the connection between the magical study of dreams and the "history of greed," specifically the links between the supernatural, dreams, and financial desires that developed in the sixteenth and seventeenth centuries. I consider "instrumental" uses of magic—mainly oneiromancy, the interpretation of dreams to predict the future—to see how pecuniary desires and the supernatural were joined.

Such mapping highlights the shifting historical meanings of money and the supernatural, which were neither predictably nor uniformly rational. Closer examination of early modern texts about these phenomena reveal that magical wealth held multiple positions in early modern culture. Even as gold's exchange value was quite considerable, it had symbolic status or meaning apart from its value. Dreams presented opportunities to early modern people to enrich themselves; the ability to read properly esoteric, supernatural, or magical signs provided early modern dreamers with the magical tools and insights necessary for the satisfaction of financial desire.

The sixteenth century was a time of heightened interest in the new meanings of money, developed in response to global trade that not only linked Europe to Asia in new ways but also instituted the "triangle trade" across the Atlantic. Alongside this economic history, there was a heightened interest in the supernatural, and the production of the *Malleus maleficarum* in 1486 (and its continued publication throughout the sixteenth century) is but one indication of this. This confluence of factors—global trade, emerging capitalist

---

[1]   Philip Goodwin, *The Mystery of Dreames, Historically Discoursed* (London: A.M., 1658), n.p.

forms, and a newly reinvigorated interest in the magical—helped to shape new understandings of the supernatural more generally. Employing supernatural mechanisms for the purpose of obtaining riches is a supremely instrumental use of magic.[2]

It seems intuitive to consider all magic to be instrumental by nature. That is, one uses magic as a tool to affect change in the world. Yet evidence from early modern Europe indicates a different conception of the magical in place. If we take the *Malleus maleficarum*, the most significant "witchhunting" guide published in early modern Europe, as an example of one of the ways people thought about magic and its relationship to instrumentality, the supernatural was not always understood as possessing the ability, or even the wish, to act as a tool that could be used to change the world. The nature of the demonic, as it was explained in Part I, Question III of the text, revolved around an abstract desire for evil: "As for his will, ... this clings immovably to evil and always sins with the sins of arrogance, envy and the highest displeasure at God's using him for His own glory contrary to the demon's will." Sinful actions and thoughts were performed for the sake of a vague evil, not for any directly obvious purpose. That is not to say that rationality was out of the question. Citing Dionysius, the author proposes that the demonic was "rational in mind, but reasoning without words" yet merely wished for the "destruction of men."[3] Evil, magic, the supernatural—these categories strained the limits of understanding, but they also did not directly translate into a tool-like use of the supernatural.

Dreams were one area in which the supernatural and the mundane could seemingly coexist, and the correct interpretation of dreams was thought to join the supernatural and the instrumental. Understanding one's dreams might in fact allow one access to knowledge that could be used to become rich. A desire to provide practical knowledge to its participants organized dream interpretation, and a link between money and dreams includes an analysis of dreams in which a treasure is revealed to the dreamer, at times through oneiric transport, through a vision, or by the direct appearance of the supernatural: a ghost, for instance.[4] According to historians like Carlo Ginzburg, dreams in the early modern period were thought of as a kind of fourth dimension: an alternate space that was diabolicized over the course of the sixteenth century by priests and peasants alike who could not fully comprehend the pre-Christian epistemes that had sustained

---

[2]   Johannes Dillinger, *Magical Treasure Hunting in Europe and North America: A History* (London: Palgrave Macmillan, 2012).

[3]   Christopher S. Mackay, *The Hammer of Witches: a Complete Translation of the Malleus Maleficarum* (Cambridge: Cambridge University Press, 2009), 124, 125.

[4]   Peter Brown, *Reading Dreams: The Interpretation of dreams from Chaucer to Shakespeare* (Oxford: Oxford University Press, 1999); Katharine Hodgkin, Michelle O'Callaghan, and Susan Wiseman, eds, *Reading the Early Modern Dream: The Terrors of The Night* (New York: Routledge, 2008).

a folk understanding of dreams and the "night battles" that took place in them for generations before the Counter Reformation transformed the dream into the devil's playground.[5] Contemporaries depicted oneiromancy as a dangerous space in which the dreamer was open to demonic influence. The *Malleus*'s author explains in Section I, Question XVI that

> the divination of dreams, is practiced in two ways. In the first, someone uses dreams to be able to track down something secret on the basis of a revelation made by evil spirits with whom explicit agreements are kept ... . In the second, someone uses dreams to learn of the future inasmuch as dreams derive from divine revelation or from a natural cause, whether internal or external. As far as such a power can extend, it will not constitute unlawful divination.[6]

Dreams that assisted in the quest for riches should be considered a type of bridge between this world and the supernatural one. If covetousness could be satisfied through the correct reading of supernatural signs, so much the better. Yet one of the central questions vexing early modern dreamers was the issue of whether or not the knowledge they gained about riches through their dreams was lawful or not.

Jean Bodin, in his *Demon-mania of Witches* (first published in 1580 and widely reissued thereafter), provided one way of divorcing illicit from lawful knowledge. Like the *Malleus*, Bodin's text served as an important manual for later thinkers seeking to understand the differences between diabolic and licit activities, and his descriptions of dreams remained important into the eighteenth century. Positing differences between diabolically generated knowledge and a legitimate form of supernatural knowledge derived from godly interventions in the human sphere, Bodin argued that there were lawful and unnatural forms of supernatural knowledge. "Natural divination," he explains, "is an anticipation of things future, past, or present—yet hidden—through knowledge of causes connected and dependent on each other, as God established them from the creation of the world. I have stated this definition in order to make certain judgment on which divination is lawful, and which divination is unlawful, or diabolical, according to the terms of the definition which we gave of a witch." Not surprisingly, Bodin drew upon his legal background to provide an explanation of why people sought to satisfy legitimate desires through illicit means: "What draws the wretched to the slippery precipice of the path of perdition, and to devote themselves to Satan is a depraved belief which they have that the Devil grants riches to the poor, pleasure to the afflicted, power to the weak, beauty to the ugly, knowledge to the

5   Carlo Ginzburg, *The Night Battles: Witchcraft and Agrarian Cults in the Sixteenth and Seventeenth Centuries* (Baltimore: Johns Hopkins University Press, 1983).

6   Mackay, *The Hammer of Witches*, 244.

ignorant, honour to the scorned, and the favour of the great."[7] Illicit knowledge, used to satisfy the covetous desires of the sinner, represented an instrumental form of magic stemming directly from the diabolical.

Bodin was not alone in these views of the power of dreams to convey important information in ways that were not counter to God's law. The mystic alchemist Paracelsus held a similar belief, writing in *Die Bücher von den unsichtbaren Krankheiten* that "God has permitted magic, and this is a sign that we may use it; it is also a sign of what we are; but we must not interpret this sign as a summons to practise magic. For if a man practises false magic, he tempts God … . And if he tempts God, woe to his soul!"[8] Later writers were equally concerned with the problem of diabolical influence in the dreamworld. Bad dreams might come from Satanic origins, good ones from divine sources. As one text from 1658 put it, "For evil Dreames, wherein the *Devil* hath his industrious dealings for mens monstrous defiling and deluding."[9] The desires themselves were not wrong, simply the method of attaining them. Magic, in other words, worked. Policing the origin of that magic and ensuring its godly legitimacy was the great task of early modern authorities.

But how could people know whether the everyday magical practices they employed to diminish the tedious grind of life in the early modern period were sanctioned or not? A lively trade in popular manuals existed to coach practitioners in magical methods that skirted the line between the diabolical and the legitimate. A survey of sixteenth- and seventeenth-century texts reveals a number of sources that fulfilled the didactic task of informing people about the virtues of licit supernatural knowledge. Interpreting dreams was a magical art, connected in many texts to the ability to read a range of esoteric sign systems: palmistry, geomancy, astrology, physiognomy, and interpreting "worry" lines on the face and the location of moles. Each provided instruction in how to "read" the language of God in these environments or how to decode the physiological signs of vice and virtue. Divination, in other words, was a multi-faceted project seeking to generate a practical form of magic that also recognized a division between the licit and the illicit. Even texts typically seen as providing a kind of diabolic workbook—Agrippa's *Of the Vanitie and Vncertainite of Artes and Sciences*, for instance—purported to teach the diabolical as a way to understand the theological better.

Oneiromancy, and how this type of knowing could help satisfy one's feelings of covetousness, provides useful ways to evaluate the declination of magic from

---

[7]   Jean Bodin, *On the Demon-Mania of Witches*, trans. Randy A. Scott (Toronto: Centre for Reformation and Renaissance Studies, 2001), 71, 155.

[8]   Paracelsus, *Selected Writings*, Bollingen Series (Princeton: Princeton University Press, 1988), 137–8.

[9]   Goodwin, *The Mystery of Dreames*, n.p.

the realm of elite practitioners to popular forms of knowledge. In the texts under consideration here, dreams were understood symbolically, but the symbols operated universally rather than individually. In the early modern period, a dream about x meant that y would happen, an understanding that was conditioned by the honing of allegorical perception. Sadly, dreams could rarely reveal practical or instrumental knowledge, but they did operate at the level of vague revelation. For example, dreams could not reveal where a treasure might be located, but they were able to alert the dreamers about the possibility of becoming rich if only they could understand the allegorical signs incorporated into their dreams.

Early modern dream literature defined a dream in multiple ways, but in all cases, dreams were understood as a realm in which the soulful, strange, and supernatural—sometimes divine, sometimes diabolical—interacted with the dreamer. One of the central tasks facing authors of dream interpretation manuals was to produce a definition of a dream that evoked its supernatural abilities but did not cross the border into the diabolical. Such knowledge was critical if one was to make use of the information that dreams could provide. Thomas Hill explains in his text *The Moste pleasaunte Arte of the Interpretacioun of Dreames* (c. 1576) that if one were to have a dream in which one were "to talk to the dead, [that would] signifyeth good and profit to follow." In other words, people had already begun to imagine that dreams had a logic to them, that they provided a symbolic system akin to our understanding of dreams. Hill distinguishes between "true" and "vain" dreams, dreams that have an otherworldly source, and those whose origins might only be located in the dreamer's diet. Nonetheless, Hill accepts the fact that "dreames seen by graue & sober persons, do signifie matters to come."[10] "Gonzalo," writing in 1641, posited a similar set of definitions that turned on the power of food to modify the dreamer's humoral balance but more importantly ascribed a simple power to the role of imagination. Gonzalo explains that "A Dreame is that which appeareth to us while we are sleeping; not by the function of the eyes, but by imagination." Dreams retained a powerful effect on the emotions long after the dreamer had awoken. The origin of this emotional power stemmed from the nature of dreams themselves, which Gonzalo adamantly described as "causes naturall," unlike those "ignorant people" who understood dreams to be "the soules of deceased persons, or Angels" who visited one in the liminal space of the dream.[11] Dreams,

---

[10]   Thomas Hill, *The Moste Pleasuante Arte of the Interpretacion of Dreames Whereunto Is Annexed Sundry Problemes with Apte Aunsweares Neare Agreeing to the m Atter, and Very Rare Examples, Not Like the Extant in the English Tongue. Gathered by the Former Auctour Thomas Hill Londoner: And Now Newly Imprinted* (London: Thomas Marsh, 1576), n.p.

[11]   Gonzalo, *The Divine Dreamer, or A Short Treatise Discovering the True Effect and Power of Dreames Confirmed by the Most Learned and Best Approved Authors : Whereunto Is Annexed the Dreame of a Young Gentleman Immediatly Before the Death of the Late Earle of Strafford* ([S.l.: s.n.], 1641), leaf B.

to be clear, could provide useful information, and that information came from some unknown but magical source.

Later texts, like the English-language update of the second-century oneiromancer Artemidorus's *Interpretation of Dreams*, published in 1644, continued in this didactic mode. This text advanced the opinion that dreams had a divine origin, and therefore the knowledge that they imparted could be considered licit: "Some are of opinion [sic] that Dreames which arise of Naturall and Carnall affection, are likewise to be interpreted; As an Vsurer to dream of gold; or any other carnall men, when they dreame of such things as their natures are prone and subject unto."[12] The mention of usury and gold here provides a useful index of the ways these manuals sought to connect dreams to the "nature" of the dreamer. Tryon claims that

> Doctors of *Morality* [have] been always advised to take notice (amongst other things) of his usual *Dreams*, there being fearce any thing that more discovers the secret bent of our minds and inclinations to *Vertue* or *Vice*, or this or that particular Evil, as *Pride*, *Covetousness*, *Sensuality* or the like, then these nocturnal sallies and reaches of the *Soul*, which are more free & undisguis'd, & with less reserve than such as are manifested when we are awake.[13]

A greedy person would dream of money; like followed like. In these definitions drawn from sixteenth- and seventeenth-century texts, dreams remained a remarkably stable way of "seeing" beyond one's own self.

Dreams revealed much and the subjects they exposed had much to say about dreamers and their world. But dreams remained spectral forms of knowledge. Even when the signs seemed clearly to indicate something, dreamers also needed to beware what remained concealed. The information in a dream was useful, but only if it could be interpreted correctly and false leads could be avoided. Dreams provided valuable information, but only to those skilled in interpretation and savvy to the ways that dreams could mislead. If one wished to use dreams as a way to become rich, for instance, the manuals could assist. Thomas Hill, whose 1576 manual is seen as an early example of the form, wrote quite clearly about the instrumental nature of oneiric divination. He explains in the preface that "If it be superstitious (gentle Reader) and therefore denied of some men, to have a foresyghte and judgemente in thinges to come, whye is not then denied to learned Physicions, skilfull warriours, weary husbandemen, and polytyke

---

12 Daldianus Artemidorus, *The Interpretation of Dreames, Digested into Five Books by That Ancient and Excellent Philosopher, Artimedorus*, ed. R. W. (London: Bernard Alsop, 1644), preface [n.p.].

13 Thomas Tryon, *A Treatise of Dreams & Visions Wherein the Causes, Natures, and Uses, of Nocturnal Representations, and the Communications Both of Good and Evil Angels, as Also Departed Souls, to Mankind* ([London: s.n., 1689]), 6–7.

Captains, to have knowledge in the Artes of divination."[14] The interpretation of dreams was a skill that should be cultivated. Goodwin took things a step further, arguing that dreams provide "profitable knowledge" of both a "certain" and a "conjectural" quality. Goodwin lards his descriptions of the "rationall soul" of the dreamer with the language of labor. Dreams were a type of "Thought-working" that created the conditions of useful knowledge: "So that by the Knowledge of Dreams, much of mans rationall soul may be certainly known."[15]

Nevertheless, the sign structure of dreams was never simple, and signs were unsurprisingly inconsistent from author to author. If we only consider those elements of the dreamscape concerned with money, riches, and financial wellbeing, we see a surprisingly wide range of signs: "to dream you see a *Small Spring*, grow into a *great lake*, denotes encrese of Riches;"[16] "To dream one is in a pleasant Meadow, signifies the possession of Riches, and the advantage of pleasure;"[17] "Have a long beard, shows Strength or Gain;" "Hear Beeasts, Signifies Gain;" "Take Bees, Signifies Profit;" "Hear a Cock crow, is Prosperity;" "See Dragons, signifies Gain."[18] Hove's *Oniropolus* (1680) explains that "To dream that thou dost give away some gold,/Doth some unpleasant News to thee unfold:/but if thou Dream, that Gold thou dost receave,/Auspicious Fortune, and good Luck thou'l have." Riches appeared again on the description of the significance of "On keeping Cattel. To Dream of keeping Beasts, to' th Rich portends/Disgrace and loss, to th' Poor it profit sends." Not to be outdone, the body appears as a sign of wealth in some circumstances: "On the Breast and Papps. To dream of Breast and Papps covered with Hair/Gain, and Profit, does thereby appear/To Men: but if a Woman have such Dream,/Loss of her Husband, is meant by the same."[19]

Especially when correlated with the astrological, money possessed a complicated set of meanings: "To dream of Money, when the Moon you find/in Aries, shews sickness, or grief of Mind:/Taurus a heavy dream: Scorpio, Theft;/Gemini of a Friend thou art bereft:/Cancer, Capricorn and Pisces also,/In all these Signs the Moon a Ghest doth shew;/In Leo Money, Virgo weariness:/Libra, Death of an enemy then guess./In Sagittarius the Dream is vain:/But Joy

---

[14] Hill, *The Moste Pleasuante Arte of the Interpretacion of Dreames*, n.p.

[15] Goodwin, *The Mystery of Dreames*, n.p.

[16] Anon., *Aristotle's Legacy: Or, His Golden Cabinet of Secrets Opened* (London: J. Blare, 1699), 16.

[17] Anon, *The Art of Courtship, Or, The School of Delight Containing Amorous Dialogues, Complemental Expressions, Poems, Letters and Discourses upon Sundry Occasions Relating to Love and Business ...* ([London]: I.M., 1686), n.p.

[18] H. Curzon, *The Universal Library: Or, Compleat Summary of Science* (London: George Sawbridge, 1712), 364–6.

[19] Frederick Hendrick van Hove, *Oniropolus, or Dreams Interpreter* (London: Tho. Dawkes, 1680), 29, 36.

in Aquarius doth remain."[20] The author of the *True Fortune-Teller*, J.S., writes that "To dream of small pieces of money, and that your gathering them off the ground, denotes want and hard labour; but to dream a good sum is put into your hand, denotes either you shall receive unexpectedly some sum of money, or that you shall be relieved by a Friend." Predictably, "To dream you have Gold and Silver in your hands, and know not how you came by it, betokens the finding of some precious thing."[21] The seventeenth-century update of Artemidorus offered a polarity, where a single sign could speak of different outcomes:

> For a man to dream he hath gold is not bad, because of the matter; as every one will say, but contrariwise it is good, as I have known by experience: but oftentimes when one hath dreamt that he hath too much, or an excesse, and ill agreeing to the [?] by reason of the fashion and figure: as to men, billiments, chains, and carquenets, in like manner: as to poore men, to dream of a crown of gold, and plate and great pieces of gold, for when any one shall have such a dream the gold signiefieth evill, not in respect of the matter, but the workmanship and figure. But if the head [?] be lost broken, or bruised, in the dream, it is loss to a woman, Losse of Rings to a man, signifes not onley the losse of them that had charge over this goods, as the wife, the tennant, etc. but also the losse of his goods, lands and possessions, or that he will not lend or give away any more, to such as he hath lent and committed charge to before times: to many, this dream hath foreteld losses of eyes, for the eyes have some agreement with the rings, because of the radiance of the stones: but when as your dreame is eyther of hosen or shoes, we must judge as of the apparel.[22]

In a similar fashion, "To dreame to buy of sorts of things which one useth, is good: to buy that which only for victual and relief, is good for the poore, but to the rich and wealthy, it signifies expences and great charge."[23]

Dreams centering on fantasies of acquisition were especially prone to dual readings. The seventeenth-century Artemidorus, for instance, explains that "For to dream to get and heap up goods, and especially fayr houshold stuffe, and well ordered, and much, or any more then we had before, is good: but not most excellent above our estate and quality, for this would be without reason, and would signifieth much hurt."[24] Tryon's interpretation of mining dreams also sought to draw the line between normal modes of acquisition and

---

[20]   Hove, *Oniropolus, or Dreams Interpreter*, 45.

[21]   J. S., *The True Fortune-teller, or, Guide to Knowledge Discovering the Whole Art of Chiromancy, Physiognomy, Metoposcopy, and Astrology* (London: E. Tracy, 1698), 105, 107.

[22]   Artemidorus, *The Interpretation of Dreames*, 64–5.

[23]   Ibid., 147.

[24]   Ibid.

morally troubling ones centered on the dark urge to possess more than one's legitimate share:

> If men did but believe those things, and diligently observe and weigh them, they would certainly be much more careful in moderating their thoughts and affections, neither would so much dote upon, or be perpetually vexed with, for, or about the Spirits of golden mountains of Ethiopia, the Dust of Guinea, the Rubish of Peru, which do give great advantage to, and powerfully attract the evil Demons, who by many of the Antients are thought to have some kind of rule over, or affinity with those hidden Mines and Treasures in the bowels and dark Caverns of the Earth, which are the Idols whom most people now adore, and over whom the evil spirit Mammon is said to be Lord president.[25]

This diversity of signification should not give us pause. Instead of seeing evidence of bad faith in these writers' works, we should instead be mindful of the ways that dreams, especially when it came to the question of money, allowed early modern dream interpreters to express their inventiveness and to fill in the connection between sign and signified in historically contingent ways. One way to read the range of signs that appear in this context is to consider that the symbolics of wealth, connected in these manuals to the bounty of nature (fields, streams, lakes, mines), to human virility and fecundity (beards, breasts), or to the vehicles of agricultural surplus (beasts, bees, fowl, cattle) continued to be organized around peasant visions of security and increase. The fact that there could be legitimate acquisition and illegitimate greed should not distract us from the larger point that the symbols of financial security often remained rooted in the logic of rural accumulation.

Others were not so sure that dream symbols, particularly when they touched on supernatural topics, could be so readily understood. Paracelsus argued in his 1537–38 text *Astronomia Magna*,

> The interpretation of dreams is a great art. Dreams are not without meaning wherever they may come from—from fantasy, from the elements, or from another inspiration. Often one can find something supernatural in them. For the spirit is never idle. If the earth gives us an inspiration—one of her gifts—and if she confers it upon us through spirit, the vision has a meaning.[26]

Moreover, Paracelsus indicates in "De occulta philosophia" that "The dreams which reveal the supernatural are promises and messages that God sends us directly; they are nothing but His angels, His ministering spirits, who usually

---

[25] Tryon, *A Treatise of Dreams & Visions*, 163–4.
[26] Paracelsus, *Astronomia Magna*, in *Selected Writings*, 134–5.

appear to us when we are in a great predicament ... ."[27] Dreaming, in other words, is a kind of supernatural contact zone. Dreams act as a messaging system not functionally different than any other type of hermetic knowledge of the natural world. Paracelsus does not really acknowledge a division between physical and metaphysical, and the dream is not any more obviously symbolic than anything else. Dreams, instead, offered a revelatory possibility, and magic offered access to the areas of experience that were typically impossible to reach. "Magic," Paracelsus argues in "De occulta philosophia," "has power to experience and fathom things which are inaccessible to human reason. For magic is a great secret wisdom, just as reason is a great public folly."[28]

Paracelsus's understanding of the supernatural is key because of his importance to the history of alchemy. If we shift our focus momentarily to alchemy, as another form of potentially illicit knowledge about the natural world that presented one with the opportunity to satisfy one's desire for riches, we see a similar set of concerns about the ability to read esoteric signs. For Paracelsus, the point of alchemy was not the production of wealth. The absence in Paracelsus's discussions of alchemy of any type of money-making desire is notable in light of the intense interest in alchemy generated by the mercantilist desires of political entities throughout central Europe. Throughout the Holy Roman Empire courts sought to employ "entrepreneurial alchemists" who could replenish their depleted coffers.[29] Yet for Paracelsus, the emphasis was on purity, not exchange, and he provides an example of the ideological cleanliness to which alchemy aspired. For the alchemist, gold was a cipher of theological meanings, not just a conduit of human exchange. Alchemy presented a challenge—but not an obviously diabolical one—to the divine order. Recent scholarship indicates that alchemy, cloaked in mercantilist thinking, was seen as a practical art and was intensely instrumentalized because it offered early modern polities a way to increase the wealth of a particular district.[30]

The trick facing rulers, always difficult of course, was the task of policing fraudulent alchemists and preventing swindlers. But in mining, too, magic could be employed to locate new sources of wealth, especially if that wealth could be generated through magic that was not directly diabolical. While it is outside of the scope of this chapter to consider the ways that treasure-finding magic like alraune-based divinations were thought to assist one in the finding of mineral wealth, we can examine how magic, mining, and alchemy could contribute to a larger concern about greed.[31] As we saw in the above passage in which Tryon

---

[27]   Paracelsus, "De occulta philosophia," in *Selected Writings*, 135–6.

[28]   Ibid., 137.

[29]   Tara E. Nummedal, *Alchemy and Authority in the Holy Roman Empire* (Chicago: University of Chicago Press, 2007).

[30]   Nummedal, *Alchemy and Authority in the Holy Roman Empire*.

[31]   Similar concerns are explored in the chapter by Johannes Dillinger in this collection.

offered his guidance in the interpretation of mining dreams, mining—and the ways mining might enter into a dreamer's vision—was a way to gain important knowledge of riches. Agrippa sought to provide insight into the ways a person could intuit the best locations to search for ore. He took up this question in a text published in the 1520s and reproduced widely thereafter, *On the vanitie vnd uncertaintie of the arts and sciences*, explaining that greed was a root cause of such searches: "By this Arte, all worldly wealth is maintained, for the greedinesse whereof suche a fantasie came in mennes braines ... /Dame nature did it hide and put/where gredie ghoostes do dwell:/And now the hurtfull yron, and/the glittering golde from hell."[32] For alchemists, too, the problem centered on the ability to describe the licit desire to obtain gold.

The lure of riches was so dangerous, and greed was such a strong emotion, that this desire led to all sorts of other terrible actions, and Paracelsus suggests in *Das zweite Buch der Grossen Wundarznei* (1536) that "Where money is the goal, envy and hatred, pride and conceit, are sure to appear—and may God protect and preserve us all from such temptations!"[33] He advances in "Liber prologi in vitam beatam" a vision of the connections between financial desires and diabolicism: "But he who loves riches sits on a shaky limb; a little breeze comes—and it enters his head to steal, to practice usury, to drive hard bargains, and other such evil practices, all of which serve only to acquire the riches of the devil and not those of God."[34] But gold and wealth were particularly significant in light of Christian bans on covetousness and avarice. Part I, Question VII of the *Malleus* explains that "as for the fact that only what has been recognized is loved, let illustrations be given at one's discretion about the gold that the greedy man loves because he understands its virtue and so on."[35]

Imagination remained a critical zone of interaction between a person and the supernatural; desire itself was the playing field upon which larger battles between good and evil were fought. Given space for expression in dreams, and allowed to develop in more tangible ways in mining and alchemy, temptations were the language through which desires found concrete expression. More than just a wish that sought fulfillment, early modern dreams about money or the satisfaction of financial desire were also a way to tap into larger forces, ones that sometimes might have skirted the division between licit and illicit powers. These early modern texts give us access to some of the ways that the non-instrumental magical world was understood. Dreams, like alchemy, acted as a contact zone, an area in which knowledge could be produced about supernatural, divine, or

---

[32] Heinrich Cornelius Agrippa von Nettesheim, *Of the vanitie and vncertaintie of artes and sciences* (London: Henry Wykes, 1569), 97–8.

[33] Paracelsus, *Das zweite Buch der Grossen Wundarznei*, in *Selected Writings*, 69.

[34] Paracelsus, "Liber prologi in vitam beatam," in *Selected Writings*, 178.

[35] Mackay, *The Hammer of Witches*, 181.

malevolent forces. Of course magic could be instrumentalized, but when it was not, then the extremes of the supernatural come into sight. And perhaps most importantly, we can glimpse the ways that the popular lore of dream interpretation—a form of knowledge that appears in these texts to be at once esoteric and democratic—gave voice to the desires of everyday people to obtain greater financial resources. Produced at a time during which the logic of a capitalist and money economy was quickly developing—and the consequent social instability that transformation entailed—it is little wonder that there was a parallel development of interest in forms of magic that could be employed to become rich. Simply put, new forms of economic exchange and the mutant logic of mercantilist constraint led people in the sixteenth and seventeenth centuries to employ magic in new and vitally instrumental ways.

Chapter 7

The Good Magicians: Treasure Hunting in Early Modern Germany

Johannes Dillinger

This study addresses the interrelation between economic behavior and specific aspects of the magical worldview prevalent in early modern Europe by comparing two elements of the early modern culture of magic that historians have rarely discussed in relation to each other: treasure hunting and the belief in witches. Even though we know now a great deal more about treasure hunting than we knew only ten years ago, this kind of magic is still understudied. Treasure hunting was by no means just another curious detail of folklore, let alone a mere footnote to the vast record concerning witchcraft. The evidence we have so far suggests that thousands of trials against treasure hunters took place in early modern Europe.[1] For practical reasons, this study will focus on source materials from the German lands, particularly those including modern Germany and Austria. Through an investigation of treasure hunting and the magic employed by treasure hunters, it argues that common people as well as legislators and judges did not identify treasure magic with witchcraft. The decisive reason for this was a concept of everyday magic: both witchcraft and the magic of treasure hunters corresponded to a specific early modern understanding of economics and economic behavior. The belief in magic was mundane insofar as it was interrelated with the economic activities of everyday life and expressed deeply rooted economic mentalities.

---

[1]   Johannes Dillinger, *Magical Treasure Hunting in Europe and North America. A History* (London: Palgrave Macmillan, 2012); Johannes Dillinger, *Auf Schatzsuche* (Freiburg im Breisgau: Herder, 2011); Yves Marie Bercé, *A la Découverte des trésors cachés du XVIe siècle à nos jours* (Paris: Perrin, 2004); Charles Beard *The Romance of Treasure Trove* (London: Sampson Low, 1933); George Hill, *Treasure Trove in Law and Practice from the Earliest Time to the Present Day* (Oxford: Clarendon, 1936); Warren Dym, *Divining Science. Treasure Hunting and Earth Science in Early Modern Germany* (Leiden: Brill, 2011). In spite of the somewhat misleading title, Dym's study is mainly about the use of the diving rod in mining.

Treasure Magic

In the early modern period, treasure hunting was clearly a magical activity, and the treasure itself was a magical object. Indeed, it was so magical it is difficult to see the treasure as an object in the modern sense of the word: for example, treasures were allegedly able to move on their own account. Very like game, treasures had to be lured, and they were capable of actively escaping the treasure hunters. In addition, they could fool treasure hunters by shape-shifting: in the very moment in which they were finally found, they turned into worthless materials, dirt and bits of wood. Only an expert magician could bring them back to their original form.[2]

Treasure magic included various ways to deal with the treasure's spirit guardians. The non-human protectors of treasures were of supreme importance for early modern treasure beliefs. Early modern Germans believed that fairies were the guardians or owners of treasures. If these mysterious beings were addressed in the proper way they could be persuaded to give away at least a part of their riches. In eastern Germany the *Drache* or *Drak* played a major role in treasure lore. A spirit that was often supposed to take the form of a flying fiery snake, the *Drache* was no medieval dragon. Rather, it was a household spirit comparable to the house or hearth spirits in snake or bird form found in Hungarian and Baltic tradition. The *Drache* was said to bring grain or money to his host. "Den Drak haben" meant in the Saxon dialect of the seventeenth century to be well off. However, a person rumored to "have a Drak" enjoyed a dubious reputation at best. The *Drache* was often said to steal the things it brought to his master from the neighbors. In the context of witch trials, the *Drache* was clearly a demonic being.[3]

---

2   Johannes Dillinger, "'Das ewige Leben und fünfzehntausend Gulden'. Schatzgräberei in Württemberg 1606–1770," in *Zauberer—Selbstmörder—Schatzsucher. Magische Kultur und behördliche Kontrolle im frühneuzeitlichen Württemberg*, ed. Johannes Dillinger (Trier: Kliomedia, 2003), 221–97, here 232–33; Dillinger, *Magical Treasure Hunting*, 58–61.

3   Dillinger, *Magical Treasure Hunting*, 66–72 and Johannes Dillinger, "Money from the Spirit World," in *Money in the German Speaking Lands*, ed. Mary Lindemann (forthcoming). See also the folkloristic account Dagmar Linhart, *Hausgeister in Franken. Zur Phänomenologie, Überlieferungsgeschichte und gelehrten Deutung bestimmter hilfreicher oder schädlicher Sagengestalten* (Dettelbach: Röll, 1995), 213–67. *Drache* is the German term for the medieval dragon that at times is said to guard a treasure, and *Drak* is its old dialect form. This fantastical monster is of course not meant here. Whether the belief in the *Drache* as a household spirit that brings money was rooted in medieval epics about dragons is open to debate, though it seems highly unlikely. For the belief in household spirits in animal form, see Yvonne Luven, *Der Kult der Hausschlange* (Cologne: Böhlau, 2001).

Demons could also be treasure guardians. The best-known example is Mephisto in Goethe's *Faust*: Faust got the jewels with which he impressed Gretchen from his demonic companion. Even though saints and angels were heavenly beings and divine agents of good, they were still supposed to guard treasures just like their very antipodes, the demons from hell. The patron saints of treasure seekers were St. Corona and, especially, St. Christopher. There were innumerable versions of the so-called St. Christopher Prayer, an often lengthy, litany-like spell that implored the popular saint to help treasure hunters find hidden riches. Alternatively, treasure seekers simply asked the saint to bring them money. Some treasure hunters reminded God himself that he was supposed to help the poor; thus he had no right to deny them the treasure. Of course, the pious treasure seekers promised in their prayers that they would reserve a part of their find for the needy or the church.[4]

We could not hope to understand early modern treasure lore, if we did not know about the belief in wandering souls. This seems to hold true for treasure beliefs in all of Europe, not just in the German speaking lands.[5] Most treasures seem to have been said to have ghosts watching over them. The idea of a wraith guarding a treasure belonged to a whole set of beliefs about the spirits of the dead doing penance or trying to fulfill certain tasks. It was commonly believed that ghosts had to walk on the earth until a task they had left unfulfilled in their lifetimes was completed or until some guilt was expiated.[6] The treasure's owner certainly had unfinished business: he had hoarded money without putting it to some proper use. Specifically, he might have failed to give the treasure to a good cause or, if the treasure had been gained by unlawful means, to return it to its rightful owner. In these circumstances, the owner of a treasure had to come back as a ghost. The discovery of the treasure was in the ghost's own interest because it was a precondition for the spirit's redemption. Because treasure hunters helped the ghost leave the visible world, treasure hunting became a godly deed and a Christian duty. The idea that the recovery of a treasure was an act of piety because it resulted in the redemption of a wandering soul was a genuine part of the motivation of many treasure hunters.[7] Of course, learned Protestantism

---

[4] Dillinger, *Magical Treasure Hunting*, 61–6, 85–90.

[5] Ibid., 72–9.

[6] Kathryn A. Edwards and Susie Speakman Sutch, *Leonarde's Ghost: Popular Piety and 'The Appearance of a Spirit' in 1628* (Kirksville, MO: Truman State University Press, 2008); Kathryn A. Edwards, *Visitations: The Haunting of an Early Modern Town* (forthcoming); Owen Davies, *The Haunted. A Social History of Ghosts*, (London: Palgrave Macmillan, 2007); David Lederer, "Ghosts in Early Modern Bavaria," in *Werewolves, Witches, and Wandering Spirits*, ed. Kathryn A. Edwards (Kirksville, MO: Truman State University Press, 2002), 25–54; Dillinger, *Magical Treasure Hunting*, 72–3.

[7] Dillinger, *Magical Treasure Hunting*, 76–9.

negated the existence of ghosts unconditionally, but this attitude seems not to have had any effect on the popular level.

In their mission to aid the spirits of the dead, treasure hunters could see themselves as playing a role in the cosmic struggle of good against evil that was in some respects comparable to that of the clergy. Demons who wanted to hinder a ghost's deliverance were supposed to show themselves to frighten treasure seekers away. If the treasure hunters wanted to succeed in their quest they had to prepare themselves: they took the sacrament, said prayers, or engaged in ritual fasting. In Switzerland, treasure seekers were even sometimes called *Schatzbeter*, i.e.,. praying men of treasure. Of course, the leadership of both churches condemned these practices as blasphemous.[8]

A *Schatzbeter* or treasure magician hardly ever worked alone. In order to fulfill their spiritual duties and the practical needs of a demanding dig, treasure hunters worked in groups. The expert magician was usually the leader of the group. Among these magicians, we encounter priests and monks or vagrants.[9] For a certain fee, they offered their knowledge and ability to find treasures and deal with demons to everyone willing to embark on a treasure hunt. The magician was usually employed by a person from the upper class, sometimes even by a prince, who organized and financed the treasure hunt. Other people from a more humble background might join the group, some of them with literally nothing to contribute to the group effort aside from their bodily strength and their willingness to do the practical work.[10] Thus, in treasure hunters' groups people from different strata of society came together who would hardly have been willing to cooperate under any other circumstances. The unifying bond between them was spiritual as well as material. The "treasure hunting parties" had features of business enterprises but they also had characteristics of religious congregations. The more emphasis was placed on the ghost's redemption as a pious deed, the more important it became to prepare for this religious task by collective prayer and devotion. A number of treasure magicians acquired the charisma of religious leaders. This parallel between priests and treasure magicians might explain why most of these magicians—in contrast to witches—were male. Treasure hunting even directly contributed to the rise of a short-lived Pietist sect in eighteenth-century Württemberg.[11] If we look beyond the scope of this collection of essays into the more or less enlightened nineteenth century, we even encounter a world-famous narrative that had a lot in common with traditional treasure tales: Joseph Smith's account of his discovery of the book

---

[8]   Stefan Jäggi, "Alraunenhändler, Schatzgräber und Schatzbeter im alten Staat Luzern des 16.–18. Jahrhunderts," *Der Geschichtsfreund* 146 (1993): 37–113; Dillinger, *Magical Treasure Hunting*, 166–71.

[9]   Dillinger, *Magical Treasure Hunting*, 153–63.

[10]   Ibid., 147–53, 163–6.

[11]   Dillinger, "Das Ewige Leben," 263–71; Dillinger, *Magical Treasure Hunting*, 168–73.

of Mormon. Indeed, the similarities between Smith's story and old treasure lore posed a major obstacle to the early Mormon mission among the educated Bürger of the German cities.[12]

Treasure magic had essentially two aims. First, treasure seekers had to divine the place where the treasure was hidden. Second, they needed to come into contact with the spirit guardians of the treasure, while simultaneously keeping them at a safe distance. Even if the treasure's guardian was a ghost who essentially wanted the treasure to be found, it was still a tricky business to deal with a wraith that was a shadowy being from the beyond and clearly perceived as frightening. For these reasons, treasure magicians were supposed to be able to have access to and be capable of handling an arsenal of magical objects. Crystals and mirrors were used as instruments of divination.[13] Some of these mirrors were known in German as *Bergspiegel* (mining mirrors): they were supposed to show objects hidden in the earth, especially mineral veins. However, the *Bergspiegel* could allegedly also reveal buried treasures. We encounter such mirrors in trials against treasure hunters time and again.[14] A seventeenth-century spellbook from Hessen described various ways to produce a magical mirror. The most simple was this: in order to turn a regular mirror into a magical object one had to write the characters ESQX on it. The mirror had to be placed secretly under the cloth covering the altar of a church, and a mass had to be said over it on three Saturdays, provided the mass was not a requiem. After each mass the magician should take the mirror and say, "I call upon you, mirror, in the name of God and in the name of my maker and in the name of the holy patriarchs and prophets and the four evangelists. You shall show me the hidden treasure wherever it may be and you shall not deceive but you shall show me the place and the spot and reveal them without any falsehood." After that incantation, the mirror had to be sprinkled with holy water and smoked with incense. Now, the mirror was ready. The treasure hunter could simply take it to a place where he thought treasure might be hidden. He turned the mirror towards the sun and said the incantation once more. The mirror would show the exact spot where the treasure could

---

[12]   Orson Hyde, *Ein Ruf aus der Wüste* (Frankfurt: Hyde, 1842); Dillinger, *Magical Treasure Hunting*, 176–9.

[13]   The magical mirror of Snow White's evil stepmother is arguably the best-known reflection of this old tradition. Her mirror was clearly a tool for divination as it could answer the apparently unanswerable question about the "fairest one," and it knew that Snow White was hiding beyond the seven mountains.

[14]   Johannes Dillinger, "Rheingold. Schätze und Schatzsucher im heutigen Rheinland-Pfalz von den Nibelungen bis zur Gegenwart," *Jahrbuch für westdeutsche Landesgeschichte* 36 (2010): 53–84; here 70; Dillinger, *Magical Treasure Hunting*, 95–6; see also the most lucid regional study Manfred Tschaikner, *Schatzgräberei in Vorarlberg und Liechtenstein* (Bludenz: Geschichtsverein, 2006), 32, 41.

be found.[15] In a way, the magician who created the *Bergspiegel* did the real work of treasure hunting—provided there were no spirit guardians to deal with—even if he was not necessarily identical with the person who actually used the mirror to unearth the treasure.

For early modern treasure hunters, divining rods were much more important than *Bergspiegel*.[16] The rise of the divining rod to the relative prominence it still enjoys began when it was used by prospectors in the rising mining areas of eastern Germany around 1500. Rods were a variant of the magic staff, and there is no reliable evidence that they were specifically used to find hidden objects prior to the second half of the fifteenth century. Only then does a Swabian spell mention hazel sticks used to find buried treasure.[17] Water-witching was also a relatively new form of magic but clearly linked to early treasure hunting. The Bavarian law against witchcraft and superstition of 1612 referred to superstitious practices associated with treasure hunting, which miners and people who dug wells also commonly used.[18] While the law did not mention the divining rod, it would be the only magical implement used by treasure hunters, miners, and people searching for hidden springs alike. The rod as the instrument of water-witches was apparently first mentioned explicitly in a short French tract on mining published in 1632.[19] By the end of the seventeenth century at the latest, the divining rod had become the universal detecting implement of Old Europe's magical culture. It was used to find virtually anything, including buried treasure.[20]

Even though a number of late medieval and early modern authorities passively ignored the use of the divining rod or even accepted it actively in mining, it

---

[15]   Hessisches Hauptstaatsarchiv Wiesbaden, 144a/36 Bd. I.

[16]   The divining rod is still used today, mostly for water-witching. I have written extensively about the history and practice of dowsing elsewhere: see Johannes Dillinger, "Dowsing from the late Middle Ages to the Twentieth Century: The Practices, Uses, and Interpretations of an Element of European Magic," *Studies in History* 28 (2012): 1–17; Johannes Dillinger and Benno Schulz, "The Divining Rod. Origins, Explanations, Uses, Thirteenth to Eighteenth Centuries," in *Heresy, Magic, and Natural Philosophy*, ed. Louise Kallestrup (forthcoming).

[17]   Gerhard Eis, *Altdeutsche Zaubersprüche* (Berlin: De Gruyter, 1964), 146–9.

[18]   Wolfgang Behringer, *Mit dem Feuer vom Leben zum Tod. Hexengesetzgebung in Bayern* (Munich: Hugendubel, 1988), 178–9.

[19]   Martine de Berterau, Baronne de Beau-Soleil, *Véritable Déclaration faite au Roy* (n.l.: n.p., 1632).

[20]   Hermann Sökeland, "Die Wünschelrute," *Zeitschrift des Vereins für Volkskunde* 13 (1903): 202–312, here 205–7, 280–87; Dillinger, *Magical Treasure Hunting*, 105. The German poet Eichendorff understood this phenomenon and took it as the basis for his Romantic poem "Wünschelrute" (Divining Rod): "Schläft ein Lied in allen Dingen, / die da träumen fort und fort. / Und die Welt hebt an zu singen, / trifft du nur das Zauberwort." ("A song sleeps in all things / which lie dreaming for ever and ever / And the world will start to sing / if you find the magic word.")

always smacked of magic and was often unambiguously condemned as such. The most important theologian and the most influential mining expert of the sixteenth century agreed with the rejection of dowsing as magic. For Martin Luther, the use of the divining rod clearly violated the first Commandment and should therefore be punished as magic.[21] Georg Agricola, arguably the father of scientific mining, knew that many prospectors used the divining rod. However, he rejected it because divining did not belong to the "natural means by which mineral veins can be discovered." Rather, it had sprung "from the conjurers' impure sources."[22] Nevertheless, only a few hardliners condemned dowsing explicitly as witchcraft.[23]

At times, the magician involved in a treasure hunt distinguished himself from anyone else mainly and essentially by the fact that he owned a spellbook. A spellbook that was supposed to contain the most effective incantations to placate spirits was often regarded as an indispensable prerequisite for a treasure hunt. In 1732, a vagrant trickster who claimed that he could get treasures made a great impression on the innkeeper of Bendern in western Austria and his guests because he had a book from which he read some mumbo-jumbo while he swung a ceremonial sword. A priest from Dornbirn foiled a treasure hunt in 1773 by depriving the treasure seekers of their magical writings.[24] In a dramatic case, two peasants who had been interested in a treasure hunt for years finally embarked on their venture when they met a student, heretofore unknown to them, who claimed that he had a copy of a grimoire allegedly written by Dr. Faust himself. This treasure hunt ended tragically with the death of three people in 1715. It became quickly known as the Christmas Eve Tragedy of Jena, and it sparked a lively discussion that may have hastened the progress of the so-called Enlightenment in Germany.[25]

If the magician did not own the book he needed, he knew how to get one in the "Black Market" for magic. We still know too little about the more or less clandestine production and sale of magical objects, especially magical writings, but there clearly was an illegal occult book trade whose networks spanned hundreds of miles. In one case during 1778 a simple farm hand from

---

[21] Martin Luther, *D. Martin Luthers Werke: Kritische Gesamtausgabe*, 120 vols (Weimar: Böhlaus Nachfolger, 1883–2009), 1:520, 3–15, 10 I/1, 590.7–591.13 (commonly cited as WA); cf. Jörg Haustein, *Martin Luthers Stellung zum Zauber- und Hexenwesen* (Stuttgart: Kohlhammer, 1990), 91–4, 98–100.

[22] Georgius Agricola, *De re metallica* (Basel: Froben, 1556) ed. Paolo Macini and Ezio Mesini (Bologna: Clueb, 2003), 26–8: "modis naturaliter venae possunt inveniri" and "ex incantatorum impuris fontibus."

[23] See for example Johann Sperling, *An virgula mercurialis agat ex occulta qualitate* (Wittenberg: Röhner, 1668).

[24] Tschaikner, *Schatzgräberei*, 28–9 and 43.

[25] Dillinger, *Magical Treasure Hunting*, 124–30.

southern Bavaria travelled to Swabia to obtain a magical book for treasure hunting supposedly written by Albertus Magnus. He got the grimoire through a middleman from the Montafon who had the book from a person in Mainz. In the middle of the eighteenth century, treasure seekers from Worms searched for a special book of spells for several months. They willingly paid for two costly trips to Frankfurt to get it. A competing treasure hunter from Worms claimed unabashedly to own a book of spells that would help him to find treasure and contained incantations for "the kings of Hell." The problem was that he needed a priest to read these incantations—and the first Catholic clergyman he asked went straight to the authorities.[26] When faced with criminal charges, a notorious treasure magician who haunted the eastern shore of Lake Constance in the early 1770s told the judge indignantly that he never used any "Doctor books." Instead, God himself and the Virgin Mary had given him two books with golden letters written by the angels that led him to every treasure.[27]

Some of the treasure magicians' books contained the litany-like spells so typical of folk magic. They called upon God and the saints, especially St. Christopher, the "patron saint" of treasure seekers, to help them find the treasure and defend them against evil spirits. Other treasure grimoires called rather openly upon these evil spirits alone to reveal treasures. A short manuscript book of spells from late seventeenth-century Hessen, for example, listed the names of four demons that supposedly could bring or show treasure. Still other magical incantations addressed ghosts and fairies or simply nondescript spirits that should be able to reveal a treasure. In 1741 the Habsburg authorities in Swabian Stockach confiscated two pieces of paper a treasure hunter had used; they were covered with planetary symbols, bits of ecclesiastical texts like INRI and Pater Noster, and a jumble of letters. This may have been used in the manner of a Ouija board to communicate with spirits.[28]

It must suffice here to hint only briefly at some other elements of the huge variety of magical objects and rituals in the treasure hunters' arsenal. In order to ward off demons, treasure hunters drew magical circles. Magical writings often gave examples of such circles and the signs and characters that had to be placed within and around them. Many treasure hunters used further protection such as amulets. The most simple, but also the most difficult of the magical rites treasure seekers had to observe, was strict ritual silence. They were not allowed to talk, let alone to laugh, during the dig. Otherwise, the treasure would vanish forever, even if they had already found it and merely needed to take it out of the ground.[29]

---

[26]   Dillinger, "Rheingold," 71.

[27]   Tschaikner, *Schatzgräberei*, 49–50.

[28]   Ibid., 28–9. A reproduction of the magical papers is in Dillinger, "Das Ewige Leben," 236–7, 258.

[29]   Dillinger, *Magical Treasure Hunting*, 108–13.

During digs, treasure hunters often combined these practices to form an impressive magical array. In 1679, during a treasure hunt in the forest near Böblingen, all of the treasure hunters wore amulets to protect them against evil spirits. The wizard who led the group had a lead tablet with magical signs on it and discovered the treasure site with a divining rod over which he had said a secret spell. When the place where the treasure was buried had been located, the sorcerer drew a magical circle with some symbols in it on the ground with a sword. He put birch twigs on the edge of the circle, apparently an original addition to the magical routine that was meant to strengthen the circle. The magician then said a lengthy conjuration in a foreign language, which he read from a bit of paper, apparently an excerpt from a spellbook. Only after this ceremony were the other treasure hunters allowed to start digging and then in strictest silence.[30]

Treasure Magic and Witchcraft

Strangely enough, treasure hunting was hardly ever condemned as witchcraft, and demonologists paid very little attention to treasure hunting.[31] English law provides the only prominent example of witchcraft linked directly to treasure hunting. In this respect, the English Witchcraft Acts differed from many Continental European laws against magic. The reason for this English peculiarity was the English legislators' interest in demonism. In contrast to English law, for example, German Imperial law never acknowledged the idea that magic implied contact with demons. The harshest English law against magic, enacted by Henry VIII in 1542, referred expressly to "dyvers and sundrie persones [who] unlawfully have devised and practised Invocacons and conjuracons of Sprites, pretending by such meanes to understande and get Knowlege for their owne lucre in what place treasure of golde and Silver shulde or might be founde or had in the earthe or other secrete places."[32] Because these persons were in contact with demons, they were as guilty of a felony as the other type of magician the Act explicitly referred to: those who used "wichecraftes inchauntement and sorceries" to kill or harm other people. All these people had to face capital punishment and forfeiture of their belongings, because when the law made treasure magic a capital offence, it accepted implicitly the old demonological argument that all magic was demonic. There could be no ameliorating circumstances or pardonable "lesser" magic. The Act itself gave the reason for this exceptionally

---

[30]   Dillinger, "Das Ewige Leben," 239–40.

[31]   Dillinger, *Magical Treasure Hunting*, 137–9.

[32]   British Witchcraft Act (1542), www.hulford.co.uk/act1542.html (accessed 24 Sept. 2012).

harsh treatment of magicians including treasure hunters: their activities caused "greate dishonor of God, Infamy and disquyetnes of the Realme." The twin aim of defending the honor of God and the good order of the state was typical for early modern legislation. For the Tudors this argument was arguably more important than for most other rulers of England. After the break with Rome and the Act of Supremacy, Henry VIII was keen to secure and to justify his newly acquired ecclesiastical power. However, the English courts did not enforce the 1542 Act, and the law was repealed shortly after Henry's death.[33] Significantly, Elizabeth I's Witchcraft Act of 1563 did not demand capital punishment for treasure hunters.[34] The Witchcraft Act of 1604 referred directly to treasure hunting only to confirm the Elizabethan Witchcraft Act, even though a new clause called for capital punishment for second offenders.[35]

German laws as a rule did not see treasure hunting as a form of witchcraft. The Imperial criminal law, Constitutio Criminalis Carolina, of 1532 focused on harm caused by magic; it mentioned neither treasure hunting nor demonic magic explicitly. In fact, when it mentioned magic, demons were never discussed. The laws of a number of German principalities outlawed treasure magic, e.g., the municipal law of Nuremberg from 1479, the Worms law of 1531, and the major codifications of the eighteenth century, such as the Bavarian Codex Maximilianeus, the Austrian Codex Theresianus, and the Prussian Allgemeines Landrecht.[36] Because treasure hunting using magic was prohibited, the authorities could confiscate the entire find, all the other (often complicated) regulations concerning treasure trove notwithstanding. The legislators apparently saw the loss of treasure as more or less sufficient punishment for treasure magic. If no treasure was found at all, the monarchical lawgiver did not take any great interest in the treasure hunt. The laws of a comparatively well-organized princedom like Württemberg ignored treasure hunting altogether.[37] The Bavarian law against witchcraft and superstition of 1612, otherwise known for its strictness, explained that treasure hunters did not make an explicit pact with the devil. No doubt, they used "superstitious arts" and even tried to invoke the devil, but they did not act in the devil's name; in contrast to witches, proper treasure hunters were not Satan's disciples.

---

[33] James Sharpe, *Witchcraft in Early Modern England* (Harlow: Longman, 2001), 15–16; Keith Thomas, *Religion and the Decline of Magic*, 4th ed. (Harmondsworth: Penguin, 1991), 292, 306.

[34] British Witchcraft Act (1563), http://www.hulford.co.uk/act1563.html (accessed 2 March 2015).

[35] British Witchcraft Act (1604), http://www.hulford.co.uk/act.html (accessed 2 March 2015).

[36] Dillinger, *Magical Treasure Hunting*, 16.

[37] Dillinger, "Das Ewige Leben," 241–2.

The punishments reflected this conclusion. According to the law of 1612, treasure seekers should be jailed or put to hard labor for a month; alternatively, they should go to the pillory or pay a fine. Second-time offenders faced double those punishments. Only third-time offenders were to be tortured in order to find out if such hardened magical recidivists did not have a pact with the devil after all.[38] In German legal practice, apart from very few exceptions, treasure magic was not treated as a capital offence. As a rule, treasure hunters were charged with superstition or fraud. They were let off with fines or some weeks of penal labor.[39] Some treasure hunters even applied officially for princely permits to search for treasure. If they promised not to use any magic to find the treasure, such permits were usually granted.[40] In the duchy of Württemberg, for example, the courts did not punish treasure hunting at all before 1683. After that, if the search had been started without official approval or if magic had been used, the treasure hunt was broken off and its participants were forced to pay the cost of the legal proceedings. Württemberg law regarded magical treasure hunting as a form of benevolent magic that did not include demonism. The culprit was to be exhorted, sent to the pillory, and expelled from his hometown; all magical items had to be destroyed. Not even this rather moderate law was put into practice. Persons who were regarded as mere "fellow travelers" or as mentally deficient got away with an exhortation. In the other cases, the punishment actually meted out by the courts consisted of infamy, a fine of less than 50 florins, a period of forced labor, or a prison sentence of up to one month; only in a rather extraordinary and late trial did the court see fit to keep a culprit under arrest for 11 months. Vagrants engaging in magic were exiled from Württemberg. Corporal punishment was not imposed on treasure seekers.[41]

Why did the courts treat treasure hunters so leniently even though treasure hunting was most obviously a very elaborate bit of magic? To be sure, most legislators and judges simply did not agree with the demonological hardliners like Peter Binsfeld who identified all magic with witchcraft.[42] Nevertheless,

---

[38]   Behringer, *Mit dem Feuer vom Leben zum Tod*, 172, 178–9, 188–90.

[39]   Tschaikner, *Schatzgräberei*, 7, 9–117, 132; Jäggi, "Alraunenhändler, Schatzgräber und Schatzbeter," 66–74; Dillinger, *Magical Treasure Hunting*, 114–46. Lenient punishments for treasure magic were certainly the rule, but there were of course exceptions. The French law of 1682 that replaced witchcraft with sacrilege as a capital offense led, when the courts applied it rigorously, to the execution of treasure magicians: see Henri Beaune, "Les Sorciers de Lyon 2," *Memoirs de l'Académie Impériale des Sciences, Arts et Belles-Lettres de Dijon* (1866/67), 69–154.

[40]   Dillinger, *Magical Treasure Hunting*, 118–21. The kings of England granted comparable permits: Ibid., 114–7.

[41]   Dillinger, "Das Ewige Leben," 241–2, 250–51, 263, 270–71.

[42]   Johannes Dillinger, *'Evil People.' A Comparative Study of Witch Hunts in Swabian Austria and the Electorate of Trier*, trans. Laura Stokes (Charlottesville: University of Virginia

there was no denying that some treasure hunters did try to come into contact with demons. In contrast to the victims of the witchcraft trials, some treasure seekers said diabolic incantations. Even if the authorities believed in the existence of ghosts—instead of regarding the apparitions of the dead as demons in disguise—negotiations with a ghost about material goods instead of a simple prayer or some legitimate church ritual for the benefit of the wandering soul were clearly unorthodox and, thus, magical. It was obvious that treasure hunters used magical objects. Even if the authorities might have been of two minds about the divining rods, the *Bergspiegel* mirrors, spellbooks, circles, mysterious signs, amulets, and ceremonial swords were blatantly magical. If one compares the evidence available in a typical trial against a treasure hunter to the evidence available in a typical witchcraft trial, the disparity is strikingly obvious. One cannot help but wonder why the treasure seeker was not condemned for witchcraft and why the alleged witch was not let off due to lack of evidence.

To be sure, most trials against treasure seekers seem to have taken place during the eighteenth century after the end of the great witch hunts; at that time even accused witches were treated more leniently. However, even the trials against treasure seekers which took place during the time of the most severe witchhunting did as a rule not charge treasure magicians with witchcraft, and the judges as well as the common people seem to have regarded treasure magic as different from witchcraft. For example, in the sixteenth and seventeenth century, suspected witches from Vorarlberg attempted to defend themselves by telling the judge that they used a spell for treasure hunting. Thus, the treasure hunt did not generate suspicions of witchcraft; rather, it was supposed to help to dispel them. The courts seem to have shared that point of view. At least the fact that the supposed witches had employed treasure magic was not used against them.[43]

There are two likely reasons why contemporaries differentiated between treasure magic and witchcraft. The first was that the belief in ghosts and the mandate to redeem wandering souls might have been stronger than the fear of witches. People were willing to give the maverick magician who engaged in treasure hunting the benefit of the doubt, even if he had a working relationship with possible demons. The second reason for the lenient treatment of treasure magicians, of course, would be that they were not supposed to harm anyone, as opposed to the witch who, by definition, practiced evil deeds (*maleficia*). Even if the treasure magician's magic might be regarded as demonic, it did not cause any harm. In theory, and generally speaking, all magic was illegal.[44] We know about the treasure hunters because they ended up in front of a judge. Nevertheless, the

---

Press, 2009), 58–60.

   [43] Tschaikner, *Schatzgräberei*, 18–19.

   [44] Wolfgang Behringer, *Witches and Witch-Hunts: A Global History* (Cambridge: Polity, 2004), 29–82.

actual practice of the courts suggested that they were not very keen to punish all kinds of magic. As a matter of fact, the large majority of magical practices do not seem to have attracted any juridical attention at all in early modern Europe. Commonplace and unspectacular variants of divination (e.g. the magical rituals used by young women to learn about their future husbands or all the practices connected with "lucky" and "unlucky" days), and protective or healing spells, provoked at best the routine criticism of clergymen or of advocates of the so-called Enlightenment; towards the nineteenth century they might have aroused the curiosity of proto-folklorists. In his books on folk "superstition," the eighteenth-century scholar Johann Georg Schmidt described literally hundreds of magical beliefs and simple rituals that do not seem ever to have had any legal consequences.[45] However, the general culpability of magic was not the point. As the trials against treasure hunters prove eloquently, not even contact with demons or the use of dangerous magical objects was necessarily a major concern for criminal tribunals and law enforcement agencies. The courts were supposed to focus their attention on witchcraft. However, just how attentive they were depended to a large degree on the amount of actual harm the community had experienced. After a devastating hailstorm, for example, rumors of witchcraft found an eager audience among court officials; without such a trigger event, they might not attract any official notice at all.[46] The belief in witchcraft brought together real harm and imaginary magic. Treasure hunting brought together real magic and imaginary benefits. What mattered were not the magical practices, but the expectation or explanation of harm and benefit. They decided what social and legal consequences the suspects had to face. It hardly mattered what kind of magic they had used, or if they had really used any magic at all.

Magical Economics

While these suggestions have some explanatory value, they do not sufficiently address why early modern Germans thought that treasure hunting was quite different from witchcraft. A third line of reasoning must consider the essentially economic character of treasure hunting which seems to set it apart

---

[45]   Johann Georg Schmidt, *Die gestriegelte Rocken-Philosophie*, 2 vols (Chemnitz: Stößeln, 1718–22; reprint Leipzig: Acta Humaniora, 1988); Margarethe Ruff, *Zauberpraktiken als Lebenshilfe* (Frankfurt: Campus, 2003); Owen Davies, *Cunning-Folk. Popular Magic in English History* (London: Hambledon, 2003).

[46]   Wolfgang Behringer, "Weather, Hunger and Fear. The Origins of the European Witch Persecution in Climate, Society and Mentality," *German History* 13 (1995): 1–27; Franz Irsigler, "Hexenverfolgungen im 15.–17. Jahrhundert," in *Hexenprozesse und ihre Gegner im trierisch-lothringen Raum*, eds Gunther Franz et al. (Weimar: Dadder, 1997), 9–24.

from witchcraft. In the remainder of this chapter, we will explore the economic element of magic further. It is not sufficient simply to regard major witch hunts as reactions to economic crises. We cannot hope to understand the belief in magic if we overlook the fact that it was part and parcel of the culture of everyday life, in which economic matters played a major role. If we study the economic aspects of beliefs about treasure hunting and witches, we might arrive at a deeper understanding of both phenomena.

At first, it does not seem productive to ask about the economic character of witchcraft. To be sure, many witches explained that the devil had promised them riches, but he always tricked them; his money turned into dirt.[47] The witches' magic did not make any economic sense in other areas either, because witches hardly ever profited from their magic. Numerous witches confessed that they had harmed not only their neighbors' cattle but also their own. The most typical and most dangerous kind of witchcraft—at least in early modern Germany—was weather magic. If the witches caused bad weather that destroyed the fields, they also ruined their own crops. From an economic point of view, witchcraft was idiotic self-harm, a kind of auto-aggression that was part and parcel of the mindless universal aggression of the witches and their demonic masters.[48]

The exception to this rule was the belief in transfer magic that surfaced in some witchcraft trials. Transfer magic had been known in antiquity; it figured among the various "superstitious" beliefs listed by Burchard of Worms.[49] Vampires were often regarded as the ghosts of magicians who had specialized in transfer magic; after their deaths they managed to transfer life force from the living to their corpses and thus gained some shadowy half life.[50] The *Drache* could be interpreted as a spirit involved in magical transfers, because it was said that the *Drache* stole the goods that it brought to its master from somebody else. The most prominent manifestation of the belief in transfer magic was probably the milk witch, who used magic in order to transfer the milk from her neighbors' cows into her own cow or even directly into her milking bucket. Transfer magic was clearly a specific kind of malevolent magic. The witches could supposedly better their own economic situations only by stealing from other people, but it was unclear from whom exactly they stole. Transfer magic just took milk (or

---

[47]   Dillinger, *Evil People*, 44–6.

[48]   Ibid., 47–52, 91–2; Ruff, *Zauberpraktiken*, 63–128; Eva Labouvie, *Zauberei und Hexenwerk* (Frankfurt: Fischer, 1991), 57–68.

[49]   Ruff, *Zauberpraktiken*, 93–4.

[50]   Romanian folklore describes rain or crop vampires who drain their neighborhoods of these resources for their own benefit: Agnes Murgoci, "The Vampire in Roumania," *Folklore* 37 (1926): 320–49, reprinted in *The Vampire*, ed. Alan Dundes (Madison: University of Wisconsin Press, 1998), 12–34, here 19–21; Éva Pócs, *Between the Living and the Dead* (Budapest: Central European University Press 1999), 67–9; Peter Mario Kreuter, *Der Vampirglaube in Südosteuropa* (Berlin: Weidler, 2001), 164–74.

honey, or grain, or fertility itself) from the neighbors or from a neighboring village. In one German folk legend, a milk witch stole a drop of milk from every farm between her own house and Rome.[51]

If we look beyond the imaginary crimes of the so-called witches to the social reality of the witch trials' victims we do find a rather elaborate economic concept. Economic behavior often played a role in the genesis of rumors of witchcraft. Persons from the economic elite were a significant minority among the victims of the witch hunts. The rich witch was typically a parvenu, a nouveau riche. She—or often he—was among those who had profited from the crises of the sixteenth and seventeenth centuries. The rich witch was not just wealthy; she had become wealthy quickly, and supposedly by harming others. The rich witches actually were economic egoists who strove to better themselves in any way possible. They were corrupt officials or their wives, political careerists, and moneylenders. The interest they charged was considered outrageous or, as a witness in a trial against a Swabian rich witch put it, "worse than Jewish." The best example was probably Dr. Diederich Flade from Trier, arguably the most prominent German witch of the sixteenth century. Flade was not only a careerist official who was notoriously open to bribes; he also lent money to a number of impoverished artisans and peasants. When people explained why they thought that Flade might be a witch they referred to his economic egoism. His economic behavior proved beyond doubt that he was a bad person.[52] And who but a bad person could be in league with the devil?

One might call this pattern the Evil People Paradigm; in a number of early modern German dialects the witches were simply known as "the evil people." This choice of words expressed exactly how people thought about witches, or rather how they identified witchcraft suspects: witches were those who, by their overtly aggressive and antisocial behavior, had seemingly proven that they were evil. Criminals and loud-mouthed beggars who had tried to bully people into giving them alms were suspected of witchcraft. Adulterers and sexual deviants were suspected of witchcraft. By far the greatest number of witchcraft suspects were people who were known to be on bad terms with their neighbors or their family, i.e., people who were overtly combative, incapable of or unwilling to live in peace. Their reputations had often been ruined by years of petty but bitter conflicts. Such "evil people" who had seemingly proven by their own behavior that they were bad, bitter, hateful, spiteful, or simply unreasonably aggressive were those

---

[51] Ruff, *Zauberpraktiken*, 91–5; Pócs, *Between the Living and the Dead*, 67–9; Éva Pócs, "Milk," in *Encyclopedia of Witchcraft*, ed. Richard Golden, 4 vols (Santa Barbara: ABC Clio, 2006), 3:765–7. Vestiges of the belief in milk witches lingered on in recent years; in southwestern Germany they even helped "explain" the success of small-town entrepreneurs. The author would like to thank Dr. Marianne Dillinger for sharing her folkloristic knowledge with him.

[52] Dillinger, *Evil People*, 84–94.

who came to mind when a community—mostly after some devastation—started looking for witches.[53] Among these "evil people" the aggressive nouveau riche were not the least. Among the aggressive behavior that seemed to characterize witches, economic behavior played a major role. Time and again, rich witches were said to be avaricious. Even though avarice did not play a major role in the learned witchcraft doctrine, in everyday life behavior understood to be covetous and acquisitive triggered suspicions of witchcraft. In the words of a witness in a South German witchcraft trial of the early seventeenth century: "Due to her great avarice people believe that she (a nouveau riche neighbor's wife) is a witch. If she is not one yet, she will certainly become one." It is important to note that suspicions of witchcraft were a kind of negative comment on the "career" of a social climber. The rich witch was not necessarily said to have become rich because she used magic and had a pact with the devil. The point was that she (or he) was supposed to use magic and to have a pact with the devil because she/he had become rich. "Evidence" for witchcraft, at least in the public eye, was the fact that a person had significantly bettered her social and economic position.[54]

In order to explain this marked dislike of people we might call pioneers of capitalism, we need to discuss pre-modern, especially agrarian, economic mentalities. Anthropology has long been familiar with the "image of limited good." People in traditional agrarian societies behaved as if all goods were only available in fixed quantities that could never be increased. Thus, one person's gain was necessarily everyone else's loss. Therefore, extraordinary economic activity and open profit seeking were discouraged and condemned as immoral. Two possible misunderstandings must be ruled out. First, the model of limited good says that people in traditional agrarian societies behave as if all goods were only available in fixed quantities. It does not suggest that people in traditional agrarian societies actively and consciously believed in the limitation of all goods as they might actively and consciously believe in life after death or in the existence of God. Second, the image of limited good shaped economic mentality, but it did not necessarily dictate everyone's actual economic behavior every time. Limited good certainly condemned profit seeking as immoral, but people do things quite often that are condemned as immoral. The image of limited good thus did not necessarily prevent profit seeking, but it put societal pressure on people who tried actively to better themselves.[55]

---

[53]　Ibid., 79–97.

[54]　Ibid., 83–95, verbatim quote 91.

[55]　George M. Foster, "Peasant Society and the Image of Limited Good," *American Anthropologist* 67 (1965): 293–315; George M. Foster, "Reply to Frans J. Schryer," *Current Anthropology* 17 (1976): 710–12; George M. Foster, "Treasure Tales and the Image of the Static Economy in a Mexican Peasant Community," *Journal of American Folklore* 77 (1964): 39–44; James Gregory, "Image of Limited Good, or Expectation of Reciprocity?" *Current Anthropology* 16 (1975): 73–84.

Certainly, the economic mentality of the limited good did not go unchallenged in the early modern period, but it clearly explains some traits of economic thinking before Adam Smith. The image of limited good helps to understand physiocratism and mercantilism, the fascination with the common good (*bonum commune*) in economic writings, the activities of guilds, the deep suspicion against strangers, and the peculiar unease the most successful urban merchants in the Netherlands seem to have felt about their own wealth—maybe even the specifically European hunger for colonial conquest as well as the workings of the village marriage market.[56]

Limited good also explains the basic structure of transfer magic: magic like that of the milk witch had to be malevolent magic because, according to the rules of limited good, the witch's gains necessarily came from another's loss. On a more abstract plain, the image of limited good clearly helps to understand the suspicion against rich witches. Everyone who bettered his economic situation, who rose on the social ladder, could only have done so at the expense of everybody else. He had violated society's norms, he had harmed (all) the others. A bad person who did something like this was most likely to be a disciple of Satan.

In traditional agrarian societies, material gain was only acceptable socially if it came from outside the society you lived in. Only if your surplus wealth came somehow from a realm "outside" of your community, that is, from a realm that was alien to your everyday experience, would your neighbors not regard your win as their loss. That meant the newly acquired good, the additional wealth, had to be explained as a gift given by spirit beings or as treasure. Thus, people who actively searched for personal gain but still shrunk from violating society's norms were likely to engage in treasure hunting.[57] Treasure hunters violated laws and used magic, but they did adhere to the norms and values of traditional society. On the village level of day-to-day contacts and tensions, social control, and presumption of limited good, treasure seeking seemed to promise gain without conflict.

This was the essential reason why contemporaries hesitated to identify treasure hunting with anti-social witchcraft even though treasure hunting was clearly a magical activity that often included contact with demons. Aggressive economic behavior triggered suspicions of witchcraft, because the rich witch had

---

[56] Winfried Schulze, "Vom Gemeinnutz zum Eigennutz," *Historische Zeitschrift* 243 (1986): 591–626; Elodie Lecuppre-Desjardin, ed., *De bono communi: The Discourse and Practice of the Common Good in the European City (13th—16th c.)* (Turnhout: Brepols, 2010); Simon Schama, *The Embarrassment of Riches* (New York: Vintage, 1987; reprint London: Harper, 2004), 289–372. If we consider the huge tracts of land given to settlers in North America and the brisk trade in land there, it is hard not to see these settlers as entrepreneurial persons relieved to be able to opt out of the "limited" European economies with all their legal, social, and cultural restrictions.

[57] Foster, "Treasure Tales"; Dillinger, *Magical Treasure Hunting*, 192–203.

blatantly violated the norms of the limited good society. She/he was the social climber whose career, according to the standards of limited good, necessarily had harmed everybody else. The rich witch was apparently ready and willing to harm others actively—by corruption or aggressive economic behavior rather than necessarily by magic—in order to satisfy her/his "avarice." The limited good certainly reduced the individual's freedom of movement in the economic sphere, but it encouraged solidarity. The rich witch seemed to have opted out of this system of solidarity.

While the rich witch seemed not only indifferent towards the economic interests of his/her neighbors but willing to violate them, the treasure hunter was the very opposite. Treasure hunting was a way of avoiding economic conflicts. The treasure hunters wanted to better their economic situations, too, but they did everything to avoid any behavior that could be interpreted as "selfish," overtly competitive, or aggressive. They would rather face ghosts and demons than alienate their neighbors. Therefore, nobody felt threatened by their magic.[58] Simply put, the rich witches' aggressive economic behavior "proved" that they were "evil," that is, likely to have a pact with the devil. The treasure hunters' economic behavior respected society's rules and their neighbors' interests, thus it proved that they were "good," that is, unlikely to have a pact with the devil, even if they used magic, but what mattered for the reputation of a person was not using magic but seeming to respect the rules of everyday life.

Limited good helps to understand why treasure magic was as a rule not interpreted as witchcraft and for what reasons social climbers were often thought to have a pact with the devil. However, we still need to address another major point in order to clarify and complete the picture. Foster's concept of limited good has at least one major blind spot, and this shortcoming becomes obvious if we look at limited good through the lens of witchcraft. When we gave the two preliminary answers to the question why treasure seekers were not condemned as witches, we observed that accusations of witchcraft were largely about damage. Demons used witches to wreak havoc, to ruin mindlessly, and with no other purpose than to destroy for destruction's sake. In Foster's concept of limited good simple destruction does not occur. His suggestion that people from largely agrarian societies understand the economy as a zero sum game does not take into account that these people rather frequently encountered the total destruction of goods. Crop failure and epidemics were facts of life that shaped the biographies of peasants. Loss and destruction caused by natural catastrophes constituted a dark background to all economic plans and hopes. It is impossible to interpret utter devastation and losses of that kind as a new distribution of resources. No one in some different part of the economy was supposed to be able to use what others had lost during pandemics or in times of starvation caused by

⁵⁸   Dillinger, *Magical Treasure Hunting*, 192–203.

bad weather. Here, loss was simply loss, for everybody, without anyone profiting from it. Christianity certainly explained disasters like crop failure or the plague and helped people cope with them, but it clearly did not suggest that material goods, health, or lives lost through such events somehow could be brought back. People influenced by the limited good mentality could see their economy as a closed system only as far as production was concerned. Limited good would explain the gain actively achieved by person X as the loss of everybody else. Such gain and such loss would be relative to each other and leave the grand total of goods unaltered. However, people in traditional societies formed by the limited good mentality did know absolute loss, like the losses caused by crop failure. Thus, the limited good could be limited even further. In the zero sum game economy, the sum could be reduced. Indeed, a major disaster might reduce the sum of the zero sum game to almost zero—or nothing at all.

This is where witchcraft came into play. Demons and witches were the agents of such depletions. With their attacks on the health and fertility of humans and livestock and especially with their weather magic, witches destroyed resources. They did not re-distribute them. Moreover, in most cases witches were not supposed to profit in any way from the destruction they caused. The cattle they had killed, the harvests they had damaged, the people they had maimed were absolute losses. The goods destroyed by witchcraft were not believed to re-surface anywhere else. They were said neither to be transferred into hell nor to become somehow the property of the devil or his followers.[59] The destruction allegedly caused by witchcraft was absolute. This circumstance, of course, made witchcraft seem even more dangerous, especially in a mental framework that saw

---

[59] Even very complex imaginings about a hidden demonic other-world where the sabbat took place or where witches went after their deaths, like the Swedish *Blåkulla*, the house of the devil in the Zauberer Jackl trials of Salzburg, or the devil's palace (*Teufelspalast*) of Ellwangen did not have such elements. The witches might have been thought to live comfortably in the otherworldly realm of the devil, even to own cattle there, but there was no connection between these goods of the demonic sphere and the goods destroyed by witchcraft in the economy of everyday life: Wolfgang Mährle, "'Oh wehe der armen seelen'. Hexenverfolgungen in der Fürstpropstei Ellwangen (1588–1694)," in *Zum Feuer verdammt. Die Hexenverfolgungen in der Grafschaft Hohenberg, der Reichsstadt Reutlingen und der Fürstpropstei Ellwangen*, ed. Johannes Dillinger et al. (Stuttgart: Steiner 1998), 325–500, here 435; Gottlieb Spitzel, *Die gebrochne Macht der Finsternüß oder Zerstörte Teuflische Bunds- und Buhl-Freundschaft mit den Menschen* (Augsburg: Göbel / Koppmayer, 1687), 177–86; Gerald Mülleder, *Zwischen Justiz und Teufel. Die Salzburger Zauberer-Jackl-Prozesse (1675–1679) und ihre Opfer* (Vienna: Lit, 2009), 236–42; Johannes Dillinger, *Kinder im Hexenprozess* (Stuttgart: Steiner, 2013), 129–31. The only exception seems to be the ninth-century episode about the magicians from Magonia mentioned by Agobard of Lyon: Agobard, "Liber contra insulam vulgi opinionem ...," in *S. Agobardi opera omnia* (Patrologia Latina 104), ed. Jacques Paul Migne (Paris: Petit-Montrouge, 1851), 147–58, here 148.

the possibilities of agricultural production as limited according to the rules of a zero sum game.

Treasure hunting—and the fear of the rich witch—belonged to a time of economic transition. This time, roughly congruent with the early modern period, witnessed the slow changes in trade, infrastructure, economic behavior, and economic mentalities one might call the rise of capitalism. Before these changes gathered momentum neither the active search for treasure nor affluent persons as witches had played a significant role in European culture. When the transition was more or less complete in the early nineteenth century, treasure hunters and witches disappeared not only out of the courtrooms, but also out of common perceptual frameworks. Treasure hunters ceased to be magicians and became (would-be) archeologists or (hobby) historians; witches were quickly reduced to the cultural stereotype of the isolated old woman. While it would be foolish to suggest that transformations in economic mentalities can explain the rise and fall of the witchcraft trials, they certainly contributed to the development of the imaginary framework in which witchcraft flourished.

Treasure hunters tapped the realm of spirits to access an outside source of economic surplus. Witches were supposed to establish contact with the realm of spirits as an outside source of destruction. The "closed" system of limited good had two openings. The first, the one of the treasure hunters, allowed access to unlimited good.[60] No one seems to have feared that at some point all the treasures would be unearthed. The other opening, the one of the witches, let in limitless destruction, and only God's intervention could keep the demons and their helpers from ruining all creation. The realization of the dream of gain without conflict and the realization of the nightmare of total loss—the magical realm of the spirits apparently offered both, and both without limits.

Conclusion

In order to secure the treasure and to ban the spirits guarding it, treasure seekers had recourse to a wide range of magical rituals and implements. Nevertheless, treasure magicians were hardly ever condemned as witches. As a rule, they were let off with only very lenient punishments. Treasure hunters were not witches not because they did not harm anyone, but because their magic positively expressed and strengthened societal values. Treasure hunts were compatible with the dominant economic mentality of the limited good. In a society that seemed to assume that the economy was a zero sum game, material gain was only acceptable socially if it came from outside society, that is, if it came as a gift

---

[60]   Patrick Mullen, "The Folk Idea of Unlimited Good in American Buried Treasure Legends," *Journal of the Folklore Institute* 25 (1978): 209–20.

from the spirit world or as a treasure. In a way, treasure magicians were "good" magicians; they did not harm anyone, either directly through malevolent magic, or indirectly through "selfish" economic behavior that violated the norms of the limited good. Witches, the evil magicians, were said to do both. Witchcraft caused harm, either by bringing total destruction or by transfer magic, that is, the magical theft some witches supposedly used to profit at the direct expense of others. In the social reality of the witchcraft trials, people who through open careerism and profit seeking had broken the rules of the limited good mentality could be accused of witchcraft. Their very "avarice" suggested that they were "evil people" in league with the devil.

Chapter 8

A Christian Warning: Bartholomaeus Anhorn, Demonology, and Divination

Jason Coy

Das Los wird geworfen in den Schoß; aber es fällt, wie der HERR will.[1]

Proverbs 16:33

In early modern Europe divination formed an integral part of learned magical beliefs and everyday magical practice. Throughout society, people turned to the occult in their attempts to foretell the future and uncover hidden knowledge. Divination took many forms, including charting the movement of heavenly bodies, watching for signs and portents, interpreting dreams, casting lots, and invoking spirits. Practiced by learned mages at court and by illiterate cunning folk in the countryside, it was ubiquitous in early modern society.[2] These practices had drawn the ire of Christian authorities since late antiquity, but in the wake of the Reformation, Protestant theologians and clergymen intensified these attacks, portraying divination as satanic in origin and an affront to proper belief in divine providence. This chapter will examine the controversy that raged over divination and fortunetelling in German-speaking Europe during the early modern period, focusing on an understudied work in the seventeenth-century demonological corpus, Bartholomaeus Anhorn's *Magiologia* (1674). Anhorn's exhaustive treatment of divination, along with the clerical discourse that informed it, shows why Protestant authorities deemed everyday magic to be such a threat and illustrates how they sought to dissuade their parishioners from engaging in occult activities. Anhorn's condemnation of popular forms of divination demonstrates that, like other seventeenth-century Protestant authorities, his attacks on divination vacillated between presenting these common magical practices as foolish delusions, as dangerous deceptions used

---

[1] "A lot is cast in the lap, but its every decision is from the LORD." (German version is from 1545 Luther Bible).

[2] For a concise introduction to early modern divination, see Edward Bever, "Divination," in *Encyclopedia of Witchcraft: The Western Tradition*, ed. Richard M. Golden, 4 vols (Santa Barbara, CA: ABC-CLIO, 2006), 1:285–7. See also Euan Cameron, *Enchanted Europe: Superstition, Reason, and Religion, 1250–1750* (Oxford: Oxford University Press, 2010), 63–9.

by Satan to ensnare wavering Christians, and as serious challenges to the clergy's monopoly on interpreting divine providence.[3]

In recent decades, scholars have greatly increased our knowledge of witchcraft and witchhunting in medieval and early modern Europe, contributing dozens of local case studies and surveys. Most research has considered witchcraft in isolation, ignoring other, more prosaic forms of occult activity, including magical healing, treasure finding, and divination.[4] According to Michael Bailey, "within the history of magic, the subjects of witchcraft and witch hunting dominate the early modern period, overshadowing in the historical imagination other forms of magical practice and concerns over superstition and improper belief that were also prevalent throughout this age."[5] While modern scholars have mostly ignored divination, many of the most famous and influential demonological treatises of the early modern period, including Heinrich Kramer's *Malleus maleficarum* (1486), Nicolas Rémy's *Daemonolatria* (1595), and Bartholomaeus Anhorn's *Magiologia* (1674), included exhaustive treatment of it. Meanwhile, many of the most important Protestant theologians of the Reformation era, including Martin Luther, Philipp Melanchthon, Jean Calvin, and Heinrich Bullinger, also devoted sermons and treatises to the topic of divination.

Clerical authorities throughout early modern Europe sought to persuade their parishioners to abandon quotidian magical activities that had long flourished in rural villages and urban settlements alike. These practices, popular throughout Europe but condemned by Christian authorities, included casting lots, uttering incantations, marking auspicious days, interpreting dreams, reading palms, scrying with crystal balls, casting horoscopes, and discerning omens. Leading scholars, including Stuart Clark and Michael Bailey, have

---

[3]   For Protestant efforts to condemn divination and judicial astrology and to advocate instead theologically acceptable forms of reading signs of divine providence in portents and biblical prophecy, see Stuart Clark, *Thinking with Demons: The Idea of Witchcraft in Early Modern Europe* (London: Oxford University Press, 1999), 318–19.

[4]   According to Michael Bailey, "the topic of early modern witchcraft and witch hunting has been examined extensively, indeed, far out of proportion to other areas of magic and superstition." See Michael D. Bailey, *Magic and Superstition in Europe: A Concise History from Antiquity to the Present* (Lanham, MD: Rowman and Littlefield, 2007), 7, 91. See also P.G. Maxwell-Stuart, ed., *The Malleus Maleficarum* (Manchester: Manchester University Press, 2007), 10–11. Recently, a series of pathbreaking studies have appeared that explore a wide range of popular magical beliefs and practices in German-speaking Europe. These include Edward Bever, *The Realities of Witchcraft and Popular Magic in Early Modern Europe: Culture, Cognition, and Everyday Life* (London: Palgrave Macmillan, 2008); Johannes Dillinger, *Magical Treasure Hunting in Europe and America: A History* (London: Palgrave Macmillan, 2011); Stephen Wilson, *The Magical Universe: Everyday Ritual and Magic in Pre-Modern Europe* (London: Hambledon, 2003).

[5]   Bailey, *Magic and Superstition in Europe*, 179.

demonstrated that the efforts of Catholic demonologists like Peter Binsfeld and Martin Del Rio to root out these sorts of folk magic stemmed from a desire to differentiate the supposedly maleficent rituals of demonic sorcery from their own, beneficent rites.[6] Protestant authorities joined their Catholic rivals in condemning divination after the Reformation, seeking to curtail magical and superstitious practices as a means of building and enforcing confessional conformity.[7] Protestant and Catholic demonologists labored to draw similar distinctions between fraudulent, or even satanically inspired, prognostication and the pious efforts of the clergy to divine God's providence from scripture and from the interpretation of daily events.

Writing at the end of the European witch hunts, the Swiss pastor Bartholomaeus Anhorn (1616–1700) contributed an exhaustive treatise to the demonological corpus as an attempt to defend these traditional attitudes towards diabolical magic and witchcraft. Born in the village of Fläsch in Graubünden, the son of a Reformed minister, Anhorn followed his father into the pulpit in 1634 after completing his studies in Basel. By 1638, he had been promoted to city minister of St. Gallen. A decade later he migrated to the Palatinate in Germany, before returning in 1661 to become a pastor in the parish of Bischoffzell in Thurgau. After taking up this position, Anhorn published several books and sermons, the most important of which was the massive compendium he wrote condemning magic, a work entitled *Magiologia: A Christian Warning Against Superstition and Sorcery.* Published in Basel in 1674, it was intended as a bulwark against the opposing threats of skeptical atheism and magical superstition, and the 1,200-page *Magiologia* argued for the reality of diabolic witchcraft and sorcery.[8] Despite the importance of Anhorn's contribution to the demonological literature, and what it reveals about clerical attitudes towards popular magic at the end of the European witch hunts, modern scholars have largely ignored him, exploring the rise of judicial restraint and elite skepticism

---

[6] See Michael D. Bailey, "The Disenchantment of Magic: Spells, Charms, and Superstition in Early European Witchcraft Literature," *American Historical Review* (April 2006): 383–404, here 386–8. See also Clark, *Thinking with Demons*, 528–30; H.C. Erik Midelfort, *Witch-Hunting in Southwestern Germany, 1562–1684: The Social and Intellectual Foundations* (Stanford: Stanford University Press, 1972).

[7] Bailey, *Magic and Superstition in Europe*, 193.

[8] Bartholomaeus Anhorn, *Magiologia: christliche Warnung für dem Aberglauben und Zauberey: darinnen gehandlet wird von dem Weissagen, Tagwellen und Zeichendeuten, von dem Bund der Zauberer mit dem Teufel, von den geheimen Geisteren, Waarsagen, Loosen und Spielen, von den Quellen, Heiss-Eisen und Wasserprob, von dem Laden im das Thal Josaphat und Bluten der ermordten Leichnam, von der Sauflerey, Verblendung und Verwandlung der Menschen in Thier ...* (Basel: Johann Heinrich Meyer, 1674).

among his contemporaries instead.[9] However, Anhorn's warnings against popular forms of prognostication and prophecy, the culmination of an extensive corpus of demonological literature from the late sixteenth and seventeenth centuries, reveal the nature of Protestant opposition to the widespread use of profane divination and the arguments clerics used in their attempts to persuade their parishioners to eschew such popular forms of everyday magic.[10]

Divination as Demonic

The most common attack that Protestant authorities like Anhorn levied against divination centered on its evidently demonic basis. For Protestant demonologists, the source of divination was ultimately "Satan, the great deceiver of the world," who had sown the seed of superstition among the pagans since the dawn of time.[11] Protestant authorities were deeply troubled that superstitious practices like augury and astrology remained so prevalent among their parishioners, despite their obvious demonic origins and God's manifest condemnation of sorcery and prognostication in Deuteronomy.[12] Protestant theologians and demonologists attacked the widespread belief in divination—reflected in folk practices and learned astrology alike—by linking these practices with

---

[9]   The biographical information provided here on Anhorn is adapted from Manfred Tschaikner, "Anhorn, Bartholomäus (1616–1700)," in Golden, *Encyclopedia of Witchcraft*, 1:39–40. Despite the prominence of Anhorn's writings, which went through several print runs during his lifetime, many recent studies of magical belief in early modern Europe, including Stuart Clark's *Thinking with Demons* and Euan Cameron's *Enchanted Europe*, for example, fail to mention Anhorn and do not include the *Magiologia* in their bibliographies. One exception to this is Ursula Brunold-Bigler, *Teufelsmacht und Hexenwerk: Lehrmeinungen und Exempel in der "Magiologia" des Bartholomäus Anhorn (1616–1700)* (Chur: Kommissionsverlag Desertina, 2003), although this work focuses amost exclusively on Anhorn's views on wichcraft.

[10]   Many Protestant theologians of Anhorn's day in Switzerland, Germany, Scotland, and England wrote comparable demonological treatises, arguing for the dangers inherent in demonically inspired sorcery and superstition. See, for example, Bernhard Albrecht, *Magia: das ist, Christlicher Bericht von der Zauberey und Hexerey* (Leipzig: Johann Albrecht Mintzeln, 1628); Joseph Glanvill, *Saducismus Triumphatus, or, Full and Plain Evidence Concerning Witches and Apparitions* (London: J. Collins, 1681); Michel Berns, *Gründliche und völlige Wiederlegung der bezauberten Welt Balthasar Beckers D. aus der heil. Schrifft gezogen: wobey zugleich unzählige curieuse Antiquitaeten* (Hamburg: Gottfried Liebzeit, 1708).

[11]   Anhorn, *Magiologia*, 19, 21.

[12]   In this passage, Anhorn replaces the simple language of the Bible with a long list of contemporary forms of prognostication (e.g., predictive astrology). For Protestant views on the demonic operation of common superstitious practices, based upon this expansive interpretation of Deuteronomy, see Clark, *Thinking with Demons*, 483–84.

the devil. Thus, Protestant theologians and demonologists sought to convince their readers that popular forms of magic, including even seemingly harmless forms of fortunetelling, were evil and satanic and to link popular forms of folk superstition to demonic *maleficium*.[13] In his massive compendium on witchcraft and magic, Bartholomaeus Anhorn sought to present magic as inherently demonic, a view derived from over a century of Protestant thought.

Protestant demonologists like Anhorn argued that since biblical sources and the writings of the church fathers had excoriated pagan forms of divination as demonic, any knowledge of the future gained through divination was derived from a satanic pact. The roots of this view, that divination and the demonic were intertwined, lay in late antiquity with St. Augustine's *De divinatione daemonum*, which taught pious Christians that all fortunetelling was based upon an implicit pact with the devil. Divination was a crucial part of Roman polytheism, prompting early Christians to excoriate the practice as a way of differentiating their new mystery religion. The views of such patristic writers, as well as the biblical condemnation of sorcery and divination in Exodus 18, prompted Christian emperors in fourth-century Rome to ban "pagan" magical practices, which increasingly came to be viewed as rooted in diabolism. Thus, in 438 the Christian emperor Theodosius II outlawed all magical rites, including divination, and this condemnation was passed on to the medieval church after the collapse of Roman authority in the West. By the High Middle Ages, these prohibitions had been codified in canon law despite the continuing popularity of folk divination within medieval society.[14] In the infamous *Malleus maleficarum*, the fifteenth-century Dominican inquisitor Heinrich Kramer was unequivocal in linking divination and demonism, asserting that "divination ... is done through the explicit and intentional invocation of evil spirits."[15]

The notion that divination was based upon a demonic pact not only survived the Reformation, but was given greater prominence in the writings of Protestant theologians and demonologists, who proved more concerned about Satan than were their Catholic rivals.[16] Martin Luther, in his lectures on Genesis delivered at Wittenberg, describes the satanic origin of prophetic dreams, seeking to dissuade the faithful from engaging in dream augury, a prominent form of divination:

> The Devil is a powerful and extraordinary spirit. He can deceive those who are awake as well as those who are asleep. Sometimes, to be sure, his dreams and

---

[13]   In these efforts, Anhorn agreed with other Protestant demonologists of his era. In fact, Calvin himself preached—also on the basis of Deuteronomy 18—that the death penalty should be used against practitioners of popular forms of magic and divination as well as their clients. See Clark, *Thinking with Demons*, 460–64.

[14]   Bailey, *Magic and Superstition in Europe*, 9–10, 53, 82, 87–9.

[15]   Maxwell-Stuart, ed., *The Malleus Maleficarum*, 55.

[16]   Bailey, *Magic and Superstition in Europe*, 196.

explanations happen to accord with the outcome, but sometimes they do not. This is because they originate from causes that are evident. For twenty or thirty years he sees the counsel and deliberations at the courts. Then he sees how the organs of the body and soul are disposed. He also sees how the princes are educated and trained, and what their dispositions and customs are. From these he draws many conclusions ... . Accordingly, Satan can very easily send dreams and later deduce an interpretation, just as he stirs up phantasies and images or inspires outbursts of rage and lust ... . Therefore, one should place no faith in such dreams.[17]

Luther's trusted associate, the reformer and humanist Philipp Melanchthon, placed more faith in divination than Luther and was deeply interested in astrology. Like other learned men of the period, he wrote sympathetically about the application of this learned art, but in a short piece on oneiromancy, the interpretation of prophetic dreams, even the credulous Melanchthon was careful to warn readers that visions of the future appearing in dreams could be either natural, godly, or inspired by the devil.[18] Later Lutheran authorities also echoed Luther's belief in the demonic nature of divination. Johann Jacob Weckers's 1575 demonological treatise, known as the *Hexenbüchlein*, for example, warned of the dangers of divination. Citing Augustine, the author contended that witchcraft and divination were intertwined and that the two shared a common link to satanic power. He argued that prognosticators, including astrologers, palm-readers, and necromancers, all drew their fortunetelling power from the devil. Necromancers, for example, "converse with the Devil ... and do wonders with blood, tear the dead from the earth, conjure the Devil in a glass, etc."[19]

---

[17]　Quoted in Jörg Haustein, "Mensch-Prediger-Professor-Autorität: Martin Luther und die Hexenverfolgung," in *Hexen und Hexenverfolgung in Thüringen: Herausgegeben von den Meininger Museen*, ed. Andrea Jakob (Meiningen: Verlag für Regionalgeschichte, 2003), 134. For a seminal work on early Lutheran attitudes regarding prophecy and divination, see Robin Bruce Barnes, *Prophecy and Gnosis: Apocalypticism in the Wake of the Lutheran Reformation* (Stanford, CA: Stanford University Press, 1988).

[18]　Philipp Melanchthon, *Träumbuch Artemidori des griechischen philosophi: darinnen Vrsprung, Vnterschied, vnd Bedeutung allerhand Träume, wie ...* (Leipzig: Christian Michael, 1677). See also Paola Zambelli, ed., *'Astrologi hallucinati': Stars and the End of the World in Luther's Time* (Berlin: Walter de Gruyter, 1986).

[19]　Johann Jacob Wecker, *Hexenbüchlein: das ist, ware entdeckung vnd erklärung, oder Declaration fürnämlicher artickel der Zauberey, vnd was von Zauberern, Vnholden, Hengsten, Nachtschaden, Schützen. Auch der Hexenhändel, art, thuon, lassen, wesen, buolschafften, artzeney, wohär sie erwachsen, vnd aller jhrer Machination. Item was Wechsel kind vnd Wütes här, vnd daruon zuo halten sey. Allen Vögten, Schultheyssen, Amptleüten oder Ampts verwaltern, vnd Regenten des weltlichen Schwedts vnd Regiments nutzlich zuo lesen / Ettwan durch den Wohlgebornen Herren, Herr Jacob Freyherr von Liechtenberg, auss jhrer gefengknuss erfaren, vnd jetzt durch ein gelerten Doctor zuosamen bracht, vnd weitleüffige beschriben etc.* (1575), in *Theatrum de veneficis* (Frankfurt: Nicolaus Basseus, 1586).

The clerical condemnation of prognostication became more strident during the bloodshed and anxiety of the Wars of Religion in the mid-seventeenth century, especially among Calvinists, as Reformed ministers like Anhorn railed against profane attempts to foretell the future. Reformed clerics had long excoriated these occult activities, beginning in 1549 when Jean Calvin himself penned an "admonition" against astrology. Calvin preached against all forms of divination, warning that profane attempts to satisfy vulgar curiosity by using augury to peer into hidden mysteries were a sign of human weakness and vanity.[20] In one fiery homily on Deuteronomy 18, for example, Calvin presented divination as incompatible with Christianity, preaching that "God says in sum, that if we want to be his people, we cannot be wrapped up in sorceries, or divinations, or enchantments, or conjurations with the dead."[21]

Following Calvin's lead, later generations of Reformed theologians also proved vocal in their opposition to both learned and popular divination. In 1571, Hyldrich Zwingli's successor in Zurich, Heinrich Bullinger, wrote a scathing condemnation of magical practices entitled *Wider die schwarzen Künste*, a work that focused heavily on divination.[22] Prompted by a spate of cases involving superstitious practices that had come before the religious authorities that year, the treatise begins by reminding readers of the distinction between divine power and what he called the "forbidden arts." Bullinger enumerates these prohibited occult practices, associating magic and witchcraft with a broad range of divinatory practices, including astrology, fortunetelling, augury, necromancy, seeking omens, and reading signs in earth, fire, and water. Bullinger warns against these so-called "black arts," arguing that they are "superstitious and forbidden things ... prohibited by God, wrong, and devilish works." For Bullinger, all divination—learned or popular—was based upon an active demonic pact, and he condemned the unholy activities of learned magicians who sought to "trap the Devil, or an oracle spirit (*pythonem*), or a fortunetelling spirit (*warsagenden geyst*) in a crystal or a bottle," activities he described as "forbidden by God, demonic,

---

[20]   Calvin, in his *Commentary on Daniel 1:17 (CCEL)*, declares: "It would be absurd, then, to attribute to God the approval of magical arts, which it is well known were severely prohibited and condemned by the law itself [a reference to Deuteronomy 18:10]. ... God abominates those magical superstitions as the works of the devil ... ."

[21]   Calvin, *Sermon on Deuteronomy 18:9–15*. See *Avertissement contre l'astrologie judiciaire*, ed. Oliver Millet (Geneva: Droz, 1985) and Mary Potter, "A Warning Against Judicial Astrology," *Calvin Theological Journal* 18 (1983): 157–89.

[22]   Heinrich Bullinger, *Wider die schwartzen Künst, Aberglaeubigs segnen, unwarhafftigs Warsagen, und andere dergleichen von Gott verbottne Künst*, first published in the collection entitled *Theatrum de veneficis* (1586). See also Bruce Gordon, "'God killed Saul': Heinrich Bullinger and Jacob Ruef on the Power of the Devil," in *Werewolves, Witches, and Wandering Spirits: Traditional Belief and Folklore in Early Modern Europe*, ed. Kathryn A. Edwards (Kirksville, MO: Truman State University Press, 2002), 155–79.

and great, heinous sins." The Swiss reformer also railed against more homespun sorts of divination, the craft of village folk practitioners, similarly associating them with diabolism: "divination, or *spiritus pythonis*, namely fortunetelling and the revelation of secrets, is also black magic (*Schwarzenkunst*), as the fortuneteller looks in a crystal or other devilish instrument to reveal hidden matters. For example, when someone has lost something, the devil-summoner (*Tüfelbschwerer*) calls upon the Devil and asks him who the thief is."[23]

A generation later, in 1608, the Calvinist theologian Abraham Scultetus, the court preacher for Elector Frederick V of the Palatinate, delivered an even more scathing condemnation of divination in a pair of sermons printed under the title *Warning against the Fortunetelling of Sorcerers and Stargazers*. Scultetus divided all magical practitioners into two "guilds," identifying the first of these as that of the fortunetellers (*Warsager*), those who reveal what will happen in the future, including "those who mark auspicious days, those who pay attention to the cries of birds, those who seek omens, those who have prophetic dreams, and those who ask questions of the dead."[24] While the other guild was composed of maleficent sorcerers and witches, Scultetus clearly states that all of these magical practitioners draw their power from Satan, using Acts 16 as evidence that their predictions come from communicating with a demonic spirit. Citing Exodus 22:20, he asserted that God has forbidden his faithful to tolerate such sorcerers amongst them, warning of the terrible punishment he would surely visit on communities who harbored these grave sinners.[25] Decrying the popularity of divination even among his own reformed parishioners, Scultetus makes contempt for magical practitioners a sign of true Christian devotion: "Whoever hates the Devil in his heart, he also hates the Devil's companions, all fortunetellers, invokers, and spellcasters."[26]

By the 1670s, as the witch hunts waned in Europe and the dire warnings of the demonologists began to meet with skepticism, the Reformed pastor Bartholomaeus Anhorn wrote a spirited defense of traditional fears of witchcraft and diabolism, an encyclopedic work that built upon centuries of demonological literature. In this 1674 book, the *Magiologia*, he built upon the intellectual foundations of earlier Catholic and Protestant theologians as well as classical, biblical, and patristic sources. Accordingly, Anhorn worked to maintain the centuries-old idea that not only witchcraft, but also all forms of magical activity were based upon demonic agency. Throughout his massive tome, Anhorn attacked divination in all its various forms, presenting it as a

---

[23]   Heinrich Bullinger, *Wider die schwartzen Künst*, n.p.

[24]   Abrahamum Scultetus, *Warnung für der Warsagerey der Zaüberer und Sterngücker/ verfast in zwoen Predigten/ so uber die letzte vier Versickel deß 47. Capitels deß Propheten Jesaiae gehalten durch* (Newstadt an der Hardt: Niclas Schrammen, 1608), 6.

[25]   Ibid., 7–8, 10.

[26]   Ibid., 15, 27.

serious threat to the Christian community and decrying its enduring popularity among his parishioners. According to Anhorn, the devil had sown the seed of such superstition among ancient pagans with great diligence, and he lamented that divination remained so prevalent in his own day, despite its clear prohibition in Deuteronomy.[27]

Anhorn was particularly concerned with the simple, everyday magic performed by his parishioners and leveled much of his ire against dream augury, a common form of divination that had troubled Protestant authorities since Luther's day. In his treatise, Anhorn consistently associated the interpretation of dreams with the diabolical. He cautioned his readers to eschew this occult practice, a form of divination used since antiquity to predict the future that was exceedingly popular in early modern Europe.[28] For Anhorn, the practice was thoroughly diabolical, and he asserted that prophetic dreams only occurred because they were "requested and acquired from the Devil with certain words and ceremonies," citing the divinatory practices of soothsayers associated with the cult of Serapis in pagan antiquity as an historical example. He reported with alarm, however, that the practice of deliberately calling upon Satan to inspire divinatory dreams had not died out with the advent of Christianity but was still common among the "superstitious people" of his own day, who summoned the Devil to send them dreams of future lovers through elaborate bedtime rituals. The theologian also warned of dreams that Satan visited "unsought and unbeknownst" upon sleepers in order to lead them into sin and superstition.[29]

Steeped in a Calvinist cosmology predicated upon the unwavering belief that divine providence governed everything, Anhorn admitted that ultimately it was God himself who permitted the devil to inflict "sinful dreams" upon pious Christians as a dire warning that they were on the wrong path.[30] Anhorn paid particular attention to the "sinful dreams" that the devil often visits upon unwary sleepers, sexual fantasies that had concerned theologians since the Middle Ages.[31] He explained how Satan inspired lurid fantasies about "strange lovers," so that when the hapless victim awakes he finds himself "stained with filth." For Anhorn, it was "a bad sign" when one experienced these sinful dreams

---

[27]   Anhorn, *Magiologia*, 21.

[28]   For the history of dream interpretation, see Edward Bever, *The Realities of Witchcraft and Popular Magic*, 107–8.

[29]   Anhorn, *Magiologia*, 29–30.

[30]   To support his point, Anhorn cites Job 33:15–17.

[31]   For the medieval church's concern with demonically inspired dreams and nocturnal emissions, see Isabel Moreira, "Dreams and Divination in Early Medieval Canonical and Narrative Sources: The Question of Clerical Control," *The Catholic Historical Review* 89 (2003): 621–42.

of lust and pleasure, since dreams are a measure of virtue, a reflection of the rigorous soul-searching demanded of early modern Calvinists.[32]

Throughout early modern Europe dream augury was a crucial part of the repertoire of early modern cunning men and wise women, professional soothsayers who interpreted clients' dreams for a fee. To refute the veracity of the auguries offered by these "dream-prophets," Anhorn marshaled an arsenal of Christian and classical sources. Citing Deuteronomy 13—"But that prophet, or dreamer of dreames, shall be slaine; because he hathe spoken to turne you away from the Lord your God"—the theologian advocated capital punishment for these "false prophets." Warning good Christians not to put their faith in dreams, he reiterated the demonic purpose of these fantasies, asserting that "No man should turn away from the word of God or heavenly truth, and be led into error and superstition, on account of dreams, which is what the Devil seeks."[33] Here Anhorn echoed the eschatology of previous Protestant demonologists, who had preached that false prophecies were the devil's chief instruments in bringing about the tribulations that the church would endure as Judgment Day drew near.[34]

Divination as Foolish Superstition

While Anhorn presented divination primarily in terms of its supposedly demonic basis, he also argued—in blatant contradiction to this notion—that all fortunetellers were fraudulent con men, bilking the credulous for their own monetary gain. While he based his contention that divination was demonic largely upon biblical and patristic sources, the idea that fortunetelling was a naïve superstition was rooted in the opinions of respected classical figures and upon medieval foundations. Many of the most respected Roman authors, revered by early modern humanists, had written scathing works critical of the simple-minded "superstition" inherent in popular forms of divination. The Roman statesman Cicero, for example, deemed profane divination to be a morally corrupting force within society and provided examples of false predictions, and Anhorn related these stories to convince his readers that fortunetelling was futile.[35] Augustine's writings on divination also influenced Anhorn, since he argued that the foreknowledge offered to diviners by demons was frequently incorrect or even willfully deceptive.[36]

---

[32]　　Anhorn, *Magiologia*, 32–3.

[33]　　Ibid., 45.

[34]　　See, for example, Andreas Engel, *Wider natur und wunderbuch* (Frankfurt am Main: Johann Collitz, 1597), cited in Clark, *Thinking with Demons*, 373.

[35]　　Bailey, *Magic and Superstition in Europe*, 19–20.

[36]　　Ibid., 55–6.

Echoing these classical and patristic sources, Calvinist authorities had long portrayed divination as fraudulent. Despite his wholesale condemnation of divination, the German Calvinist minister, Abraham Scultetus, writing at the start of the seventeenth century, conceded that many popular forms of folk prognostication did not rely upon an explicit demonic pact. Rather than excusing these practices, however, he ridiculed them as the product of superstitious ignorance. He drew distinctions based upon the motivations of superstitious countryfolk: "Such people think that if one means well, it is not bad if one believes this or that, but they are misled and drawn day by day and farther and farther from the true God, from faith, prayer, and hope." He also poked fun at professional practitioners who dupe the credulous, wondering how "those who make astrological almanacs and draw star charts do not laugh when they meet each other in the street, since they both disguise such lies." Accordingly, Scultetus warned his flock to avoid the snare of divination, arguing that common forms of fortunetelling like astrology were "futile" and "godless" and proclaiming that "every godfearing heart should beware of it."[37]

Like Scultetus, Bartholomaeus Anhorn viewed popular forms of divination as foolish superstitions that undermined proper Christian piety. Complaining in apocalyptic terms that such "superstition … is in these last days very common," the theologian condemned both popular and learned forms of prognostication prevalent among his parishioners. Throughout his discussion of popular forms of augury, Anhorn repeatedly attacks such "superstitions" on two fronts, asserting that they are derived from pagan origins and are also a tool of the devil, meant to deceive mankind. Thus, he refutes the superstitious reading of "portents" in flights or calls of birds, calling it an "atrocity" and asserting its pagan origins. Subsequently, he describes how the calls of owls and ravens are popularly believed to be bad omens, warning that Satan uses these superstitions to deceive people just as he used the snake to tempt Eve.[38] Horrified by the dizzying variety of common superstitions that regulated the everyday life of his parishioners, Anhorn asserted that these customs were derived from "Jews, Pagans, and Turks" and had corrupted Christianity. In his condemnation of dream-prophecy, Anhorn was eclectic in his use of evidence, relying upon patristic authorities and pagan philosophers alike to buttress his portrayal of common superstitions as fraudulent.[39] Here the Swiss demonologist recounted stories of false prophesies in ancient Rome and Persia and quoted classical authorities like Diogenes and Cicero, who doubted the prophetic meaning of dreams, in order to convince his readers that "only fools place stock in dreams."[40]

---

[37] Scultetus, *Warnung für der Warsagerey der Zaüberer und Sterngücker*, 29.
[38] Anhorn, *Magiologia*, 150, 153.
[39] Ibid., 160–63.
[40] Ibid., 39–40, 43.

Anhorn complained most vehemently about learned astrologers, who scrutinized the movement of the Sun, Moon, and stars, excoriating all who sought to unlock the secrets of the heavens for this "vain superstition." Anhorn, like other early modern authorities, did not define "superstition" in the modern sense, as irrational belief, but rather as supernatural beliefs that deviated from Christian orthodoxy.[41] Before the eighteenth century, an essentially medieval definition of superstition, as "deformed or misdirected worship," persisted among Protestant and Catholic theologians alike, producing anxiety about the liminal zone of common magical practices—healing spells, charms, and divination—situated between orthodox religious rites and forbidden demonic incantations. Thus, for Protestant demonologists, whose faith was grounded on the unshakable conviction that divine providence governed all, common superstitions like palmistry or astrology represented the idle "fantasy" (*Einbildung*) that something exists that is contrary to nature and the order established by God.[42] On these grounds, Anhorn conflated astronomy (*Sternengesetze*) and astrology (*Sternenlehre*), warning that these "two sisters" are equally deleterious to religion and the service of God.[43] Lumping together the esoteric astrological hermeticism of the so-called "Platonic school" and the astronomical observations of Tycho Brahe in a similar fashion, he declared that any belief that the stars influence events on earth is "superstitious, heathenish, and godless."[44] As with dream-prophesy, he fluctuated between presenting the activities of these "stargazers" (*Sternen-guker*) as deceitful and demonic, calling such prognostication "a manifest deception of the irksome Devil."[45] To prove the diabolical origins of astrology, Anhorn informed his readers that Augustine had long ago proclaimed that astrologers often form pacts with Satan and depicted astronomers as idolaters who follow the gods of the pagans, namely, planets/gods like Jupiter and Mars.[46]

---

[41] See Michael D. Bailey, "The Disenchantment of Magic," 388–9, 402. See also Clark, *Thinking with Demons*, 485–7.

[42] Anhorn, *Magiologia*, 19–20. At the core of Calvinist religiosity was an ironclad belief in the absolute sovereignty of an omnipotent, omniscient God: that God's eternal will governed the universe, ruling out any sort of random chance in worldly affairs. According to Calvin, in Chapter 16 of the *Institutes of the Christian Religion*, "God is deemed omnipotent, not because he can act though he may cease or be idle, or because by a general instinct he continues the order of nature previously appointed; but because, governing heaven and earth by his providence, he so overrules all things that nothing happens without his counsel."

[43] Anhorn, *Magiologia*, 173–4.

[44] Ibid., 181–2.

[45] Ibid., 184, 217.

[46] Ibid., 196–7, 200, 209. To support this point, Anhorn cites several "learned Christians," including William Perkins (an English Puritan and author of *A resolution to the countryman prooving it utterly unlawfull to buy or use our yeerely prognostications* [1618] and also *A discourse*

Divination as Challenge to God and Clergy

Vacillating between presenting divination as demonic or fraudulent throughout the *Magiologia*, Anhorn's main objection to occult prognostication was that it ignored God's evident providence and prescience. A strict Calvinist, Anhorn adhered to a stark brand of theocentricity, arguing that since God's providence governed everything in the universe, pious Christians should focus all of their efforts on seeking signs of the divine will.[47] He even provided his readers with guidance on how to conduct such "pious" prognostication—seeking signs of God's mercy or his wrath in nature—interpreting the great comet of 1665 as a clear "warning from God."[48] Like other Protestant theologians, Anhorn recognized that a belief in astral destiny and divine omnipotence were mutually exclusive, and he warned his readers that belief in astrology would overturn all order and "rob the Christian religion itself of its most solid foundation": the assurance that earthly calamities like war and pestilence are punishments for sin and "come from God alone."[49] Thus, godly people should flee from astrology, since like other forms of divination it is both false and "contrived by the Devil," and strive instead to search the world around them for signs of God's providence.[50]

---

of the damned art of witchcraft: so farre forth as it is revealed in the Scriptures and manifest by true experience ... [1608]), Calvin, and Luther, as well as a Jesuit, Cornelius a Lapide.

[47] For the efforts of early modern Protestant theologians to read earthly events and wonders as "sermons" from God and as escatological portents, see Clark, *Thinking with Demons*, 365–6, 372. Also see Philip M. Soergel, *Miracles and the Protestant Imagination: The Evangelical Wonder Book in Reformation Germany* (Oxford: Oxford University Press, 2012).

[48] Anhorn, *Magiologia*, 165–66. Here Anhorn uses scripture, namely the "red in the morning" passage in Matthew 16:3, to support his interpretation of natural phenomena as potential expressions of God's will.

[49] Ibid., 212, 222. For the clash between judicial astrology's assertion of "astral destiny" and Calvinist providential theocentrism in seventeenth-century England, see Clark, *Thinking with Demons*, 189–90.

[50] Anhorn, *Magiologia*, 164–5, 183, 229–30. According to Calvin, in Chapter 8 of the *Institutes of the Christian Religion*, "if God's providence shines in the heart of the faithful person, not only will he be delivered from the fear and distress by which he had previously been oppressed, but he will be freed of all doubt. For as we rightly fear fortune, so we also have good reason to dare boldly to entrust ourselves to God." See John Calvin, *Institutes of the Christian Religion: 1541 French Edition*, trans. Elsie Anne McKee (Grand Rapids, MI: William B. Eerdmans Publishing, 2009), 454. For a lucid discussion of Protestant notions of divine providence, see Cameron, *Enchanted Europe*, 211–16. Note that many prominent Catholic demonologists shared this view. See, for example, the first German translation of Nicolas Rémy, *Daemonolatria, das ist, Von Unholden und Zauber Geistern* (Franckfurt am Main: Cratandro Palthenio, 1598), 437–46, 452–3, where the French jurist refutes divination on the basis of God's prescience.

Anhorn's views were by no means new, and his work built upon centuries of Christian theology. Since late antiquity, the Catholic Church had opposed divination not only because it was deemed a pagan survival based upon a demonic pact, but also because predicting the future presupposed a notion of fate that clashed with Christian dogma regarding free will and the untrammeled omnipotence of God.[51] Citing Augustine, the medieval inquisitor Heinrich Kramer was unequivocal about the role of divination in turning Christians from proper faith in God's prominence in his notorious *Malleus maleficarum*:

> When we know not what we should do, we have this one refuge, that we should turn our eyes to Thee. And without doubt God will not fail us in our need. To this effect also S. Augustine speaks: Whosoever observes any divinations or auguries, or attends to or consents to such as observe them, or gives credit to such by following after their works, or goes into their houses, or introduces them into his own house, or asks questions of them, let him know that he has perverted the Christian faith and his baptism and is a pagan and apostate and enemy of God, and runs grave danger of the eternal wrath of God, unless he is corrected by ecclesiastical penances and is reconciled with God.[52]

After the Reformation, Protestant, and especially Calvinist, theologians re-emphasized this point, complaining that divination undermined proper faith in God's providence and challenged the clergy's exclusive role in interpreting signs of the divine will. While Protestant condemnations of divination appeared in the early years of the Reformation, they reached a fever pitch during the late seventeenth century, as Protestants sought to create a religious community purified of Catholic and pagan "superstition," efforts that informed Anhorn's treatment of everyday magic.

Although the measured humanist and reformer, Philip Melanchthon, who studied both astronomy and astrology at the University of Heidelberg, saw no inherent contradiction in a belief in divine omnipotence and astral influence, most Lutheran authorities viewed astrology as a direct challenge to faith in God's plan.[53] For example, in his 1575 *Hexenbüchlein*, Johann Jacob Wecker asserts the cosmological incompatibility of a belief in the efficacy of divination and faith in divine providence:

---

51   Bailey, *Magic and Superstition in Europe*, 89.

52   Maxwell-Stuart, ed., *The Malleus Maleficarum*, 219.

53   Philip Melanchthon, *Orations on Philosophy and Education*, ed. Sachiko Kusukawa (Cambridge: Cambridge University Press, 1999), xxi, 94. Melanchthon sought to integrate astrology and theology by reasoning that the orderly movement of the heavens revealed God's plan for the universe.

It is certain and true that no one can say what the future holds, except the one God and those he commands through the Holy Spirit. If the warlock, witch, or sorcerer babbles on about the future, it is a sham; if they happen to guess correctly, then it all comes from the Devil, who in the end never tells the truth, except when it pleases him. Often those who are ill or who have lost something, run off to the warlocks, witches, and sorcerers seeking their help and advice, therein they are cheated and are not aided.[54]

Calvinist authorities proved even more strident in their opposition to fortunetelling. Thus, the Genevan reformer Jean Calvin provided a lucid refutation of divination on theological grounds, arguing that it was impossible to divine the future through occult means, because only God, in his omniscience, could know what the future holds. For Calvin, divination was:

an abomination before God, and so likewise are Soothsayers. It is a question of whether it is possible for a man to foretell of things, for it is God's office to foreknow things to come; how then may it belong to the devil? It is certain, as Isaiah says [Is. 41:23], that idols foresee nothing. And as for Satan, he must always needs be the father of lies and deceive all who ask counsel of him. Yet notwithstanding all this, God does now and then allow Satan to tell of things to come, and this is for the hardening of those who will not obey the truth ... . Yet it is true that soothsayers lie more often than not, and by that means our Lord deludes them that seek counsel with Satan after this fashion. And let us not think it strange, though enchanters, soothsayers, and such other like do now and then tell of things to come; for it is God's just patience so that they may be plunged into error more deeply. For as much as they would be willingly deceived, he lets them be so, that they may perish. Thus you see why the law was made concerning people who prophesy of things to come.[55]

The Reformed theologian Heinrich Bullinger also feared that belief in divination would undermine proper faith in divine providence: "God, our Lord and Father, is the ruler and governor of all creation, visible and invisible. He guides and preserves mankind in his all-powerful governance, providence, and ordinance." Thus, according to Bullinger, "the many magi, *mathmatici*, stargazers, planet preachers, and prognosticators, who seek to predict future things through the movement of this or that star and thereby to subject a man's

---

[54]   Wecker, *Hexenbüchlein*.

[55]   *The Sermons of M. John Calvin upon the Fifth Booke of Moses called Deuteronomie* (London, 1583; rprt. Carlisle, Pa., 1987), 668–71, as quoted in Alan Charles Kors and Edward Peters, eds., *Witchcraft in Europe, 400–1700: A Documentary History*, 2nd ed. (Philadelphia: University of Pennsylvania Press, 2000), 268.

life, soul, and all his goods to the stars," committed a grave sin. Through these occult activities, he argued, "the people are led astray into blasphemy and forget God's providence, his prescience, and his governance and arrange all their actions according to the stars, signs, and planets and thus trifle with idolatry and heathenism." Bullinger even argues that belief in astrology could be used to justify immoral behavior: "Many also would gloss over their sin and shame in an unchristian manner, saying, 'I must engage in whoring and adultery, because I am a child of Venus.' You are much more a child of the Devil, blinded by the forbidden art of stargazing."[56]

In his 1608 sermon on superstition and divination, the Calvinist court preacher Abraham Scultetus warned his parishioners about the vengeance God would visit on those who dabbled in seemingly harmless folk divination. Reminding his flock of the biblical story of the destruction of Judah during the reign of Manassas, who "paid attention to the cries of birds and portents and consulted fortunetellers and soothsayers," Scultetus warned that "God punishes sorcery with everlasting death." Condemning folk magic, he complained that in his own day "people commit such terrible idolatry, therefore it is no wonder that they are so frightfully punished by God."[57] Rather than rely on astrology, which he derided as "entirely false and godless," Scultetus encouraged his parishioners to turn to prayer instead:

> if you would like to get wind of the future, if times of plenty or scarcity are coming, if pestilence, or war, or peace will come, do not go to your almanac, but to your little room, and pray, appeal to God ... who cared for you before you were even born. If you want to know how it will go for you and yours, don't go to the stargazer; do not allow him to chart your horoscope. Go instead to the Word and ask King David, who will give you a good answer in the 34th and 37th Psalm.[58]

Building upon these theological foundations, Bartholomaeus Anhorn echoed the view that divination represented a threat to proper belief in divine providence, but he also presented it as a dangerous challenge to clerical authority. The Swiss demonologist saved his most scathing condemnation for the so-called "new prophets," popular seers who "predicted remarkable world events."[59] In dealing with their homespun visions and prophecies, often presented as visitations by God or his angels, he sought to associate these phenomena with satanic influence and to present them as challenges to biblical and clerical authority. For Anhorn, these demonically inspired false prophecies stood as inversionary challenges

---

56   Bullinger, *Wider die schwartzen Künst.*
57   Scultetus, *Warnung für der Warsagerey der Zaüberer und Sterngücker*, 8–9.
58   Ibid., 31.
59   Anhorn, *Magiologia*, 23–4.

to biblical prophecy and the manifest signs of divine providence perceived by the faithful in daily events.[60] As in other arguments, Anhorn cited numerous historical examples of the diabolical nature of false prophecies, beginning with classical antiquity. According to the Swiss demonologist, "among the heathens, the Devil had a better chance of assaulting them with devilish visions, giving the appearance that these came from their gods, since they had fallen into all manner of blasphemy, superstition, and grave sins."[61] Thus, he claimed that from the time of Romulus the auguries of the Romans, omens sought in the entrails of birds, were used by Satan to keep the pagans in his thrall.

Anhorn also cast aspersions upon the revelations of a plethora of religious visionaries. Thus, he asserted that the revelations of "the great lie-prophet (*Lugen-Prophet*) and founder of the blasphemous Turkish error, Mahomet," were demonically inspired, an attempt by the devil to lead the "Saracens" into blasphemy and damnation.[62] Pagans and infidels were not the only ones deceived by such demonic delusions, however, and Anhorn asserted that "riotous Anabaptists," like Thomas Müntzer during the Peasants' War and Jan of Leiden during the Münster Rebellion, had been "carrying out all their mischief under the name ... of godly visions, visitations, and revelations."[63] Here Anhorn cited examples from antiquity and the Old Testament, as well as more contemporary ones, in his attempts to prove that the prophesies of religious nonconformists like the Anabaptists were all demonically inspired.[64] A tireless defender of orthodoxy, Anhorn sought throughout the *Magiologia* to link the ecstatic religious enthusiasm of religious dissenters like the Anabaptists and Quakers with the diabolical, presenting sober Calvinist theodicy as a bulwark against Satan's machinations.[65] Citing scripture, Anhorn warned his pious readers that "the deceit of these apparitions" would grow more frequent as the end of this troubled world drew nearer, since "the Devil's rage against the holy light of the Gospel is very fierce, and he seeks to swindle the people with insignificant,

---

[60] For the theological understanding of the diabolical as a nightmarish inversion of Christian orthodoxy, see Clark, *Thinking with Demons*, 83–5.

[61] Anhorn, *Magiologia*, 74.

[62] Ibid., 75–6.

[63] Ibid., 79.

[64] Ibid., 30–31.

[65] Calvin excoriated the religious "fanatics" of his own day in print, deriding them as dangerous "Libertines" in his writings: see, for instance, his *Sermon on Deuteronomy 13:2–5*. In linking contemporary religious enthusiasm with satanic inspiration, Anhorn also mirrored the stance of prominent Anglican demonologists of the 1660s–80s, like Glanvil and Boyle, who sought to present Anglican orthodoxy as a defense against both the "atheism" of materialists like Hobbes and the dangerous, demonic enthusiasm of the Quakers. Clark, *Thinking with Demons*, 300.

fraudulent, and false apparitions and visions, using [them] for the diminishment of the Word of God and to confound the true ministry."[66]

Anhorn was particularly concerned with the prophetic visions reported in his own backyard, claiming that "there is no nation under the Sun in which so many fantastic prophets can be found, as among the Germans, especially since the start of the Bohemian war in 1618."[67] Decrying the influence of recent "superstitious prophets," like the Rosicrucian mystic Philipp Ziegler, the Württemberg folk prophet Hanns Keil, and the Lutheran visionary Johann Warner, Anhorn emphasized the threat that they posed to the authority of the clergy. Warning of the danger popular visions posed to Christian order, he asserted that

> the originators of these visions and apparitions are mostly not reputable, since they ... call for faithful repentance with all earnestness, but do so without an honest heart. They have no calling to the ministry and do not understand that they have allowed themselves to become the instruments of Satan, who raises these visions and apparitions above the Word of God, in order to sow error and superstition among the simple-minded and heedless populace through these means. The cunning of Satan is great: whosoever does not perceive this is entirely blind.[68]

Thus, while the Swiss folk prophet Peter Wieland claimed in the 1620s that God had appeared to him, warning of a coming punishment, his words sounded to Anhorn like the "scratching of the very claws of Satan."[69] Anhorn asserted that the visions that these folk prophets believed or claimed came from God really emanated either from the hallucinations induced by melancholia or directly from Satan, appearing to them in the guise of an angel. Likewise, those who looked for signs in the movement of the sun, moon, or stars were deluded: for Anhorn, the only legitimate signs the godly should seek were signs of God's mercy or his wrath.[70] The faithful should place their trust only in the prophesies of the Old and New Testament and signs of divine providence apparent in everyday life, interpreted by the ordained ministers called to that task by God.[71]

For Anhorn, the threat posed by popular prophets, who usually presented their visions as angelic messages and in Christian moralistic terms, was that they distracted people from God's word and led them into error, thereby serving as "instruments" of Satan.[72] Thus, a folk prophet named Andreas Haberfeld had "prophesied wondrous things that would occur in 1624," including the

---

[66] Anhorn, *Magiologia*, 80.

[67] Ibid., 88.

[68] Ibid., 80–81.

[69] Ibid., 82–3.

[70] Ibid., 164–5.

[71] Ibid., 109–10.

[72] Ibid., 124–5, 129–30, 133.

"rebuilding of Jerusalem and the arrival of the reign of the Holy Spirit," but according to Anhorn, "the time is past, and these prophecies have not come to pass, and Satan rages in his terrible wrath ever more ... seducing the entire world." Noting that popular visionaries like Haberfeld invariably confronted their local pastors, Anhorn postulated that Satan used these seers to persuade the community that demonic "visions and prophesies are more believable than the preaching of the clergy."[73] As evidence of this "danger," he cited a story from a Bavarian chronicle of a female folk prophet who appeared in Augsburg in 1516. For the Swiss Calvinist, the tale demonstrated the superstition inherent in the Catholic cult of saints and the threat folk prophets posed to the clerical monopoly on interpreting divine providence. According to Anhorn, this false prophet had not only seduced the simple-minded common folk, but also learned Catholic clerics, having

> spoken eloquent lies to the people: that she did not eat or drink but lived in a heavenly way through the Holy Spirit, that God and the Angels appeared to her at night and revealed secret things to her. It was not only the unlearned, common folk who believed her, but also the local magistrates and learned theologians ... and took her for a demi-goddess and a living saint, and used her words in sermons, disputations, and printed books.[74]

Conclusion

Bartholomaeus Anhorn and other Protestant theologians and demonologists attacked the widespread belief in divination—folk practices and learned astrology alike—for three main reasons. First, they argued that since biblical sources and the writings of the church fathers had excoriated pagan forms of divination as demonic, any knowledge of the future gained through divination was derived from a satanic pact. Secondly, these religious thinkers argued, on the basis of both classical and patristic sources, that all fortunetellers were frauds, making a living by swindling their customers, a notion in tension with the previous critique. Finally, Protestant, and especially Calvinist, theologians like Anhorn complained that divination undermined proper faith in God's providence and challenged the clergy's exclusive role in interpreting signs of the divine will. While Protestant condemnations of divination had already appeared in the earliest years of the Reformation, this discourse became more strident amid the religious strife of the late seventeenth century.

---

[73] Ibid., 121, 131.

[74] Ibid., 135, 138.

In his classic work on the shifting relationship between magic and religion in early modern England, Keith Thomas asserted that although the medieval Catholic Church "drew a firm line between religion and superstition, their concept of 'superstition' always had a certain elasticity about it ... . In general, the ceremonies of which it disapproved were 'superstitious'; those which it accepted were not."[75] The same is true of the Protestant clergy's treatment of divination and prophecy: readings of providence sanctioned by the Calvinist clergy were licit, whereas popular divinatory practices or ecstatic prophecies that threatened orthodox theology or clerical authority were not. Accordingly, they sought both to suppress dangerous forms of popular divination and to counter challenges to the clergy's monopoly on sanctioning individual interpretations of divine providence. For Bartholomaeus Anhorn, the belief that events could be foretold through magical means outside of Scriptural prophecy threatened the very notion of the omnipotent "almighty Creator" he espoused, threatening at once the theocentricy that was the foundation of Calvinist spirituality and the calling of the clergy to interpret the divine will.

---

[75]   See Keith Thomas, *Religion and the Decline of Magic: Studies in Popular Beliefs in Sixteenth- and Seventeenth-Century England* (Oxford: Oxford University Press, 1997; first ed., 1971), 48.

Chapter 9

The "Antidemons" of Calvinism: Ghosts, Demons, and Traditional Belief in the House of François Perrault

Kathryn A. Edwards

In September 1612 a spirit appeared in the Mâcon home of Huguenot minister François Perrault. Behaving much like the modern image of a poltergeist, it pulled the blankets off Perrault's wife, threw books from shelves, and caused as much commotion as a "charivari."[1] The spirit eventually manifested as a man who proceeded to tell Perrault and his neighbors about secret events, bemoan his current circumstances, and discuss theology. News quickly spread; the Perrault house entertained both Catholic and Calvinist observers from around the region, and stories about it were reported in Switzerland and even England.[2] The spirit's departure was as sudden as its arrival. On 22 December Perrault's neighbors found a large viper leaving his house, and after parading it around the town and announcing that it was the "Devil that came out of the Ministers house,"[3] they killed it. No further disturbances occurred. Perrault would leave Mâcon a dozen years later and live the remainder of his long life—he was approximately 80 when he died—in the region around Gex in western Switzerland.[4]

---

[1]   François Perreaud [sic], *L'Antidemon de Mascon ou la Relation pure et simple des principales choses qui ont esté faites et dites par un demon ...* 2nd ed. (Geneva: P. Chouët, 1656) is the most commonly cited early edition, although the first edition appeared in 1653. The modern French edition used here is *L'Antidemon de Mascon, ou histoire particulière et veritable de ce qu'un demon a fait et dit à mascon en la maison du sieur François Perrault*, ed. Philibert Le Duc and Alexandre-Gaspard Perrault de Jotemps (Bourg-en-Bresse: Milliet-Bottier, 1853), 24. All translations are my own unless otherwise noted.

[2]   It was first translated into English as *The Devill of Mascon. Or, A true Relation of the chiefe things which an uncleane Spirit did, and said at Mascon in Burgundy, in the House of Me Francis Perreaud Minister of the Reformed Church in the same Towne* (Oxford: Henry Hall, 1658) under the sponsorship of Robert Boyle. It went through four English editions by 1658, and there was even a Welsh edition in 1681.

[3]   Perrault, *L'Antidemon*, 31.

[4]   For biographical information, see the note at the end of Perrault, *L'Antidemon*, 177–90. Also Théodore Claparède, *Histoire des églises réformées du Pays de Gex* (Geneva:

Perrault's notoriety has faded only slightly over the centuries, at least among those who specialize in the history of witchcraft and the supernatural. The "devil of Mâcon" is one of the more famous apparition accounts from seventeenth-century France, and modern scholars frequently depict it as an example of the confessionalization of demonology occurring in that era. Yet the events in 1612 should also be understood from a perspective in which confessionalization only plays a small role. Most of the individuals involved in the case saw François Perrault's house as haunted and, as such, brought assumptions about familial relationships and the dead to the case and their attempts to accept or expel the spirit. The "devil of Mâcon" was only a devil if the prescriptions of strict Huguenot ministers are accepted, but even accounts by those who were convinced Calvinists and felt their belief to be embattled show some ambivalence when making that orthodox pronouncement. The haunting of François Perrault's house thus illustrates tensions within Calvinist theology and the complexities of reconciling the Reformed religion with existing beliefs, folklore, and social networks. It exposes the negotiations at the heart of everyday religious and magical experience and the extent to which communities lived with and even accepted ambiguities that trouble many modern scholars.

The durability of such cultural models and ambiguities is even more striking in Perrault's case because of the strength and length of his ties to the Reformed community. François Perrault came from three generations of Reformed ministers; his grandfather, Pierre, had forsaken a noble heritage, gone to Gex, and become a pastor in 1537; according to François himself, both Calvin and Farel were present at the ecclesiastical assembly where Pierre became a pastor.[5] Pierre's son, Abel, married another minister's daughter, and François was born to them in 1577. François completed his studies in Berne, and when he moved to Mâcon in 1611, he had been a minister for eight years and had just married his second wife, who was both noble and Reformed. The haunting began mere months after they arrived. Perrault would be the Huguenot minister in Mâcon for at least 13 years, and he was instrumental in getting that community a house of worship near the city. After that time he returned to his family's homeland in the bailliage of Gex, approximately 20 miles from Geneva. In Gex he served as a minister in multiple churches and remained active in that role until close to his death. François Perrault could not have better credentials as an orthodox Huguenot.

Cherbuliez, 1856), 342–3; P.G. Maxwell-Stuart, "Rational Superstition: The Writings of Protestant Demonologists," in *Religion and Superstition in Reformation Europe*, eds Helen Parish and William G. Naphy (New York: Manchester University Press, 2002), 170–82.

[5]   Perrault does not mention the disputes Calvin and Farel had with Berne's authorities that would lead to Calvin being chased from the city in 1538. In the *Antidemon* any tensions within Reformed communities, such as the Zwinglian/Calvinist disputes underlying Calvin's expulsion, are ignored.

Despite this ministerial lineage, the tensions underlying Perrault's story and his treatment of the haunting immediately appear in the way the news about the spirit was transmitted. According to Perrault, he first described these events in 1613 as part of two conjoined treatises. The first, *Démonologie ou traitté des sorciers*, outlined a general demonology that followed standard Calvinist teachings, especially those sketched by the lawyer and theologian Lambert Daneau in *A Dialogue of Witches*.[6] The second, *L'Antidemon de Mascon*, depicted in detail the events that occurred in Perrault's home and his inspiration for writing these treatises. Perrault was well aware that the demon's affinity for his household and his own inability to expel it played into the hands of Catholic polemicists in a region where the League had dominated until the very end of the French Wars of Religion, and the Huguenot population was fighting to exercise their right to religious assembly.[7] He specifically singled out "sieur Marcelin," a Capuchin preacher, who he saw as slandering him among his neighbors and in a book Marcelin had published.[8] Despite this concern, Perrault's manuscript did not appear in print until 1653. Perhaps it circulated in manuscript, a situation where Perrault could more tightly control its readership? Perhaps he thought better of increasing the notoriety of a Huguenot community that was just reasserting its right to exist in Mâcon? Perhaps he wrote to make sense out of an experience beyond any he ever expected during his life and ministry but which touched on fundamental aspects of both? Whatever the case, Perrault's account reflects the attitudes and actions of a man trying to reconcile disparate intellectual, theological, and pastoral demands.

Perrault's education, family background, and spiritual sensibilities gave him clear guidelines for how to understand such spirits. Protestant polemics on such

---

[6]   *Démonologie ou traitté des sorciers* is the title in the first edition. By the second, expanded edition in 1656, it was entitled *La Demonologie ou Discours en général touchant l'existence, puissance et impuissance des demons et des sorciers ...* (Geneva: P. Chouët, 1656). There were multiple editions of Daneau's work through the sixteenth century, including the first, in Latin: *De veneficis, quos olim sortilegos, nunc autem vulgo sortiarios vocant: dialogus, in quo quae de hoc argumento quaeri solent breviter et commode explicantur tractatus ...* (Geneva: E. Vignon, 1574). There was also an English translation of 1575: *A Dialogue of Witches*, translation attributed to Thomas Twyne (London: 1575).

[7]   Perrault, *L'Antidemon*, 61–2. Perrault's nineteenth-century editor depicts the *Antidemon* as a tract he wrote as an old man, a point that contracts Perrault himself. For information regarding possession cases in early modern France and their use as confessional polemic, see Sarah Ferber, *Demonic Possession and Exorcism in Early Modern France* (New York: Routledge, 2004).

[8]   Perrault, *L'Antidemon*, 53. Here Perrault claims his account is "naïf, simple et veritable," claiming for himself the legitimacy that simplicity carries with it and, by implication, challenging Marcelin's contrived, false narrative. The same sentiments are expressed on p. 54.

diverse subjects as purgatory and everyday magical practices labeled spirits as illusions or demons, come to test the faithful by God's permission. To assume that an apparition had some other foundation was to accept common ideas that even Catholic authorities found unpalatable and unorthodox: for example, that spirits of the dead could linger for several days after death. Any interaction with the dead was impossible, and customs such as asking questions of the dead or praying for the dead were actively dangerous because the principles underlying those practices undermined the true, Scriptural faith.[9] Ministers were well aware of the power such practices held, and Calvin himself sympathized with the human need for "solace" and to show "love to the dead," although he also described this custom as "indulg[ing] our love."[10] Spirits who claimed to be the beloved dead were preying on this love.

For a Reformed minister, such spirits must be demons. By the time of the Mâcon apparition, Calvinists like Perrault could also turn to several detailed treatises in both French and Latin—the two languages Perrault would have likely known best—discussing how to understand and interact with such demonic manifestations. Particularly likely to have influenced Perrault, or at least reflecting a shared Calvinist perspective, was Lambert Daneau's book on witchcraft and magic, and such texts frequently discussed spirits of all kinds.[11] A widely recognized polemicist and theologian, Daneau stressed the links between witches, demons, and spiritual visitations, although he left an opening for an additional explanation when he stated that, while judging such events, people should consider "whether 'the accomplishing and trueth thereof, plainly repugneth against the course of nature.'"[12] In addition, Perrault was also versed in the work of other polemicists on these topics; he explicitly mentions the writings of the French jurist and witch hunter, Pierre de Lancre.[13] If he felt that a source

---

[9]   *Institutes*, book 3, section 6.

[10]   Ibid., section 10.

[11]   Daneau, *A Dialogue of Witches*.

[12]   Daneau, *Dialogue of witches*, sig. Gviiv, as quoted in Stuart Clark, *Thinking with Demons: The Idea of Witchcraft in Early Modern Europe* (New York: Oxford University Press, 1997), 172.

[13]   Perrault, *L'Antidemon*, 56. The standard edition is Pierre de Lancre, *Tableau de l'inconstance des mauvais anges et démons, où il est amplement traicté des sorciers et de la sorcellerie ...* (Paris: J. Berjon, 1612); the most recent French edition appeared in 2000 and was edited by Nicole Jacques-Chaquin. For an authoritative English edition and translation, see Gerhild Scholz Williams, ed., *On the Inconstancy of Witches: Pierre De Lancre's* Tableau de l'inconstance des mauvais anges et demons *(1612)*, trans. Harriet Stone and Gerhild Scholz Williams (Turnhout: Brepols, 2006). It is also possible that Perrault was familiar with Catholic authors who wrote on such topics, such as Henri Boguet (based in the neighboring Franche-Comté) or Jean Bodin, since Protestant and Catholic theorists writing about witches and demons were often surprisingly ecumenical in their reading.

focusing on ghosts and demons specifically might be necessary, the Protestant Ludwig Lavater's famous treatise on ghosts had appeared in Latin in 1570 and in French in 1571; both editions were printed in Geneva and influenced many Protestant treatments of spirits.[14]

Perrault's own description and analysis of his house's haunting echoed the natural-historical and theological perspectives found in these authors and would have been acceptable for a Huguenot minister. When it came to classic natural-historical, empirical, and experiential methods of observation and interpretation, Perrault was ready to apply them and was far from credulous. Moreover, he denied that he was conducting these studies from "curiosity," a potentially dangerous and theologically fraught category.[15] For Perrault, applying these methods was besting the devil in his own playing field, given that, as Perrault noted in his *Démonologie,* the devil is a great "naturalist":

> To begin with, all those things one believes are evil spirits turn out to have entirely natural explanations; for example, if one hears the slightest noise in the house, the cause may be a mouse, a cat, a dog, a hen, a wall, a beam, or a joist which is cracking or contracting or moving because it is too dry or too damp; or perhaps the wind is blowing and shifting something we cannot see; or the eyes of certain animals; glow-worms in the night; dead wood. There are torches and lights commonly known as will-o'-the-wisp, which appear in the night, wandering from place to place and are seen especially near marshes, ponds and streams, or near cemeteries and gallows. The common folks believe these are evil spirits, but learned men think that, on the contrary, they are exhalations which have risen from the ground as far as the lowest region of the air, where they are lit by antiperistalsis. For while they rise, they are pushed back by the cold which is in the middle region of the air, and at this time they appear as hoppers, seeking places which are sloping downwards, and water, which is their contrary element.[16]

---

[14] Ludwig Lavater, *De Spectris, lemuribus et magnis atque insolitis fragoribus variisque praesagitionibus quae plerunque obitum hominum, magnasque clades mutationesque imperiorum praecedunt* (Geneva, 1570) and *Trois livres des apparitions des esprits, fantosmes, prodiges et accidens merveilleux qui précèdent souventesfois la mort de quelque personnage renommé ou un grand changement ès choses de ce monde* (Geneva, 1571). Peter Martyr Vermigli wrote three questions on this subject and their resolutions that were printed with the Geneva editions; this addition shows the importance the Genevan religious establishment attached to their topic and, because his responses echo Lavater, the support they gave to Lavater's arguments.

[15] Perrault, *L'Antidemon*, 54. He cites both Catholic and Protestant authorities in support of his conclusions, although he does so briefly, making it impossible to tell to what extent he was engaged in their arguments. See, for example, Ibid., 56.

[16] Perrault makes this argument forcefully in the *Démonologie*, as quoted in P.G. Maxwell-Stuart, *The Occult in Early Modern Europe* (New York: St. Martin's Press, 1999). Also see Perrault, *Demonologie*, 2nd ed. (1656), 76.

To apply both everyday observation and natural philosophy to the analysis of such practices was both common-sensical and spiritually necessary for Perrault: God's true manifestations, as unlikely as they were, and the devil's tests could not be fully appreciated if a person ignored the reason and knowledge God gave him. Perrault's attempts to observe and rationalize the events in his house within a natural framework thus appear from the beginning of his account. The spirit's activities first affected Perrault's wife and a maid while he was out of town, and when he investigated their reports, he resolved to be "neither too ready to believe that which they [the women] said, nor too doubtful." After all, women were notoriously unreliable witnesses. As events progressed, the spirit reported long conversations with Perrault's brother and other acquaintances; rather than presume the legitimacy of these reports, Perrault sent for human confirmation.[17] He and his neighbors ran to check the places where the spirit's voice originated for evidence of any tampering or human assistance, and he noted how frequently people could assume that the wind, creaking steps, and other natural noises could have "supernatural" explanations. In so doing, he applied theories of respiration as understood at the time, describing the ways in which natural objects could expel "breath" or exchange energy. He stressed his eyewitness' experience of the spirit's actions both within his house and in Mâcon itself. One example of such analysis occurred near the end of the spirit's visitation. At that time Perrault went to the shop of Abraham Lullier, a goldsmith, who described losing a tool and a ring despite carefully organizing his shop.[18] After spending some time trying to find them, Lullier gave up. At that moment both came falling out of the sky in front of him, and Perrault saw the two affected objects. When combined with Lullier's trustworthiness at other times, their presence served as concrete evidence that Lullier was truthful in this case, too.[19] While not all ministers might have been so persistent, Perrault was perfectly orthodox in such studies.

Perrault's treatment of and response to the apparition also contained many elements that could easily fit into a modern scholar's checklist of appropriate pastoral behavior and theology for an early seventeenth-century Calvinist minister. When the disturbances first started in Perrault's house, he presumed that they were designed to prevent Huguenots from the more ready and public "exercise [of] their religion," something for which Perrault had been working actively in the 11 months since he had moved to Mâcon.[20] Although the spirit remained invisible through most of its time in Perrault's house, and only its noises, damage, and harassment made its presence known, in mid-November

---

[17]   Perrault, *L'Antidemon*, 25, 31–3.

[18]   Lullier and another member of Perrault's congregation were responsible for getting the house for Perrault: Ami Bost, *Histoire de l'Église protestante de Mâcon* (Mâcon: A. Ruel, 1977), 203. Spirits also seem to have plagued Lullier and his shop.

[19]   Perrault, *L'Antidemon*, 50–52.

[20]   Ibid., 61.

1612, the spirit began speaking coherently and in a masculine voice with Perrault. Immediately Perrault labeled it "Satan" and claimed protection as a "servant of God."[21] The spirit's existence and lewd and profane behavior were signs of its demonic roots, just as they were that of the myriad other spirits Perrault learned were littering the region.[22] Throughout the visitation Perrault tried many orthodox ways of disconcerting the spirit—ignoring it, observing it, discussing it with his neighbors—but he portrayed prayer as the most successful option, although he rarely described people praying in his home. Despite this mixed treatment, Perrault discussed the efficacy of prayer, noting that they only needed to get on their knees as if they were going to pray for the spirit to say, "While you're praying, I'm going out onto the street." Although providence appears rarely in Perrault's account, it was a key component to his understanding of events, as might be expected. He attributes the spirit's inability to cause any real damage, especially to his study, to God's will: "He didn't even have permission to break or tear apart a single leaf of my books, to break a single glass, nor to extinguish the lamp, which remained lit all night on my table during that time."[23] In his conclusion, Perrault reaffirmed the centrality of God when he described everything as having occurred through God's will and God's mercy as having sustained him.[24]

Yet, like other famous Protestant demonologists, Perrault's actions and attitudes as he himself records them can be ambiguously orthodox. Despite Perrault's prominence in his local church and his appropriate opinions, his actions during, responses to, and interpretations of his haunted house are far from the clearly confessionalized prescriptions that are so often treated as truisms when discussing interactions between the natural and the supernatural, the living and the dead, in early modern Europe. Although Perrault had a successful and prominent career as a Huguenot pastor, his perspectives in this case show him willing to accommodate and even to conform to local concerns and networks and to have internalized traditional folkloric attitudes towards spirits. He is Huguenot, but that identity is far from all that defines him. Perrault cannot escape and apparently does not want to escape from a psychology and an environment that leads him to more ambiguous and accommodating positions.

Two examples of such an approach were Perrault's inclusivity and investigative style. Like many such events in early modern France, the haunting of Perrault's house quickly became a community sensation, and people came throughout the day and well into the night hoping to encounter the spirit. Perrault himself told

---

[21]　Ibid., 28.

[22]　For example, see Ibid., 41. The First President of the Chambéry parlement, M. Favre, was the victim of a similar haunting soon after that of Perrault: Frédéric Delacroix, "Les Procès de sorcellerie au XVIIe siècle," *La Nouvelle revue* 84 (1893): 528–48, here pp. 532–3.

[23]　Perrault, *L'Antidemon*, 46, 52.

[24]　Ibid., 65.

both the Huguenot Elders and other influential people in Mâcon soon after the spirit arrived and invited them to observe for themselves; early in the haunting he decided that it would be better if he reported it to the town authorities than if they heard about it from others.[25] In fact, he apparently welcomed or at least accepted their visits, even when they lasted into the early hours of the morning. One of his frequent guests was Simon Meissonier, who helped Perrault explore the locations from where they heard the spirit's voice coming.[26] Perrault was also careful to identify and legitimize those who came to his home; while he accepted that they came to engage with or analyze the spirit rather than pray to expel it—and he treated these deviations from orthodox Reformed procedures in a remarkably matter-of-fact way—he wanted to make it clear that it was not just anyone who was loitering around his house. For example, Perrault started his account of his conversations with Abraham Lullier, the goldsmith who provided him with several pieces of evidence about the spirit, by comparing Lullier favorably with another goldsmith who came from Geneva and "deceived" a woman in Mâcon.[27] Even though the spirit was in Perrault's house for less than four months, Perrault suggests that many other individuals, such as the First President at Chambéry, used his household as a center of investigations into activities by spirits from throughout the Mâconnais. Perrault's own story suggests that his home became one of the main gathering spots in Mâcon, even taking on some of the characteristics of a village pub when visitors played games with the spirit and sat drinking until all hours.

Moreover, these advisors and acquaintances included some of the leading Catholics in the region. Mâcon is generally depicted as highly confessionalized, with its few Huguenots scrambling to retain a foothold, and there are good reasons for this assessment.[28] Perrault frames his account by telling about the battles he and his fellow Huguenot had to get Mâcon's town council to recognize their royal privilege to have a place of worship closer to the town, and the need to oppose Catholic lies is his justification for writing the story

---

25   Ibid., 27. Also see pp. 28 and 53–5 for lists of individuals who had to see about the spirit for themselves.

26   Ibid., 39.

27   Ibid., 40.

28   Although circumstances had been peaceful enough in 1595 to allow a debate between the minister of the Huguenot church, Théophile Cassegrain, and the Franciscan Father Humbelot, Mâcon's Huguenots had some difficulties in practicing their faith in the 1610s. For example, their place of worship was at Hurigny, about six miles from Mâcon. This situation only improved in 1620, when they gained a church approximately half a mile from town. See Emile Magnien, *Histoire de Mâcon et du Mâconnais* (Mâcon: des Amis du Musée de Mâcon, 1971); Pierre Goujon, ed., *Histoire de Mâcon* (Toulouse: Privat, 2000), ch. 4 and 5; Bost, *Histoire de l'Église protestante*, 43–200. See especially ibid., 203–12, for Perrault's time as minister in Mâcon.

down.[29] Yet events tell a more complex tale. As soon as Perrault was convinced that something supernatural was occurring in his home, he consulted with well-reputed men, some of whom also happened to be committed Catholics. These same Catholics visited with him every day during the haunting and stayed with him almost every night until at least midnight, if not later. One, in fact, took notes and prepared a transcript that was lost, at least according to legend, during the Revolution.[30] The sense Perrault gives is that they were doing so as support or through curiosity, not as confessional one-upmanship. Like his Catholic colleagues, Perrault was upset to hear the spirit say the Our Father inaccurately and was especially incensed at the way it treated anything to do with the Holy Spirit.[31] He also seemed more confused than offended when the spirit asked him to send for the priest of St. Etienne parish so that it could confess its sins; he even recounted phlegmatically that the spirit urged them to remind the priest to bring his holy water. Although it is unclear to what extent Perrault was actually happy to accept these requests—the spirit's comments about holy water happen in a paragraph where Perrault censured it for mocking religion—they did not inspire any diatribe about false religion. In fact, Perrault's matter-of-fact tone occurred anywhere Catholic beliefs were discussed, although he was willing to curse and condemn the spirit as a demon in other circumstances. As such, his consultation of Catholics and acceptance of Catholic practices is a sign of the ambiguous nature of confessional boundaries even in a region where Huguenot–Catholic relations were tense. It reinforces arguments for the importance of community cohesion and suggests that, in certain cases, there may have been more shared than disputed values and belief.[32]

Perrault's ways of treating the spirit reflected similar ambiguities. Traditional Calvinist ways of dealing with the demonic possession of a person or an object

---

[29] Perrault, *L'Antidemon*, 21–2 (a colloquy at Bourg de Couches that he had to attend to defend Huguenot rights in Mâcon) and 61–2 (the town councilors and "notables" of Mâcon challenged royal commissioners about the Huguenots' right to worship closer to town).

[30] Ibid., 52, describes the tests that sieur Tornuz put the demon through when it would not speak to him. On p. 55, Perrault notes that sieurs Tornuz and Chambre made an accurate and detailed report to the archbishop and prepared memoirs of what they experienced. Pages 53–4 contain a detailed list of all the Catholics who came to see him, visits that Perrault attributed in part to the lies Friar Marcellin spread.

[31] Ibid., 27, 34.

[32] For other examples of confessional coexistence in early modern France, see Gregory Hanlon, *Confession and Community in Seventeenth-Century France: Catholic and Protestant Coexistence in Aquitaine* (Philadelphia: University of Pennsylvania Press, 1993); Ole Peter Grell and Robert W. Scribner, eds, *Tolerance and Intolerance in the European Reformation* (New York: Cambridge University Press, 1996); Benjamin Kaplan, *Divided by Faith: Religious Conflict and the Practice of Toleration in Early Modern Europe* (Cambridge, MA: Belknap Press of Harvard University Press, 2007).

were through faith, prayer, good behavior, and the avoidance of wicked thoughts; active expulsion, such as exorcism, was seen as compelling God and, therefore, illegitimate. While Perrault never suggested that he considered exorcism, he actually seemed to take little action at all. There is little sense that he or his household prayed or did much that was overtly religious during the time the spirit was there. One of the few mentions of prayer occurs just after Perrault notes that God has not allowed the spirit to damage anything in his house; by linking this statement to one that he prayed and will continue to pray on his knees "all the days of his life" he makes his prayer seem slightly coercive—as if the spirit's inability to do damage was a natural, almost obligatory, divine reward for such prayer.[33] His actions seem especially odd in light of standard arguments for why books such as *The Antidemon of Mascon* were produced. Omitting any sense of God's prerogatives would undermine the book's pedagogical value and proof of Protestant righteousness. Moreover, while Perrault certainly cursed the demon, there was nothing particularly Calvinist about that—in fact, cursing demons was a standard part of exorcism—and if he intended such curses to drive away the spirit, they were ineffective.[34] In one episode, Perrault demanded the spirit say Latin prayers as a test, something which could be seen as a condemnation of Catholic approaches to the supernatural except that Perrault himself gave the prayers their Latin names and specifically asked for the spirit to say them in Latin. As such, it was merely conforming to Perrault's orders.[35] Even Perrault's interpretation of changes in the spirit's demeanor and actions was confessionally neutral. Like many theologians of his time, Perrault argued that apparitions of the devil were becoming more frequent and desperate and that, in individual cases of possession, the devil grew progressively more violent as the world neared the end times because he felt that he would be unsuccessful in corrupting the faith of those he had tormented.[36]

Although Perrault's methods of interpreting the spirit may have been ambiguous, his analysis of it was far from it: the spirit was a demon. Given that certainty, it seems odd that Perrault was more puzzled than shocked when members of his household and community had a friendly relationship with the spirit. Perrault's wife shared his convictions and determined to put her trust in God and remain in her home, but their maid acted as if no such resolve was necessary. She talked with and teased the spirit, even relaying its message that the Perrault's two young boys were perfectly safe while they slept in the room

---

[33]   Perrault, *L'Antidemon*, 53. Perrault and his wife both expressed their trust in God's protection by refusing to flee the home (p. 45).

[34]   Ibid., 42.

[35]   P.G. Maxwell-Stuart, *The Occult in Early Modern Europe* (New York: St. Martin's Press, 1999), 49; Perrault, *L'Antidemon*, 34.

[36]   Perrault, *L'Antidemon*, 51 and 55–6. There is nothing particularly Calvinist in this perspective; many Catholic theologians would agree.

next to that where most of the noise occurred. This chambermaid eventually left—Perrault was quite vague about why—and the spirit continually harassed her replacement.[37] Perrault also noted that there were some people with whom the spirit particularly liked to speak, and these individuals had no fear; rather, they joked with it and even mocked its knowledge and behaviors. For example, Michel Repay came with his father to see the spirit almost every evening. One evening he laughed that, "Father, I swear that he [the spirit] speaks exactly like my mother!"[38] Perrault reported that claim in the same tone as he did when he noted that he had a kitchen. For Perrault and his household, though, it seemed that the most surprising welcome was that of the dog, who was notorious for barking at the least noise but never feared nor barked at the spirit.[39]

Even Perrault moved beyond the role of passive recipient of demonic speech into that of active conversationalist with the spirit. He discussed news and nature with it, and at times he seemed almost sympathetic. In that sympathy Perrault slid dangerously close to the patterns of human charity and spiritual manipulation about which Calvin warned.[40] As Perrault concludes his description of the haunting, he notes that he could tell many more stories about many more individuals but that discretion prevents him. The impression was that the spirit was willing to talk in almost all circumstances, it was surprising when it did not respond to guests, and Perrault was present for most, if not all, of the gossip.[41] The possibilities for extraordinary knowledge were especially tempting, and Perrault was almost smug when the spirit confirmed that Perrault's father was poisoned and another neighbor was murdered, things that "everyone believed to be true." When the spirit bemoaned the fate of France's Huguenots, Perrault's concern implied that he saw this prediction as another example of truthful, hidden knowledge. He seemed almost pleased to have supernatural confirmation of his worries.[42]

Particularly striking in Perrault's account was a series of apparitions when the spirit seemed to be two different spirits. Near the end of the haunting, the spirit changed its demeanor and spoke slowly and piteously. It begged for assistance in making a will because it had to leave soon for Chambéry to have a legal case judged there and it was afraid of dying on the road. Perrault seemed taken aback, but its meekness and religious sensibility clearly touched him deeply. The spirit asked Perrault to bring a notary, which he did, and the spirit itemized his

---

[37]   Ibid., 37, 62–3.

[38]   Ibid., 38.

[39]   Ibid., 34.

[40]   See Calvin's warning about the dangers of sympathy leading the simple to "grosser superstitions" in the *Institutes*, book 3, chapter 5.

[41]   Perrault, *L'Antidemon*, 46.

[42]   Ibid., 33, 43.

bequests.[43] Perrault is unclear who was the spirit's main heir or how Perrault contacted him, but Perrault went to the bother of doing so, implying that it was believed to be or considered worth treating like an actual person. Perrault also described the spirit as speaking very politely to him, claiming not to know him, and "pretending a short time after" to be a different spirit than the one who had visited until that time, which it described as "his master." This behavior on the spirit's part ended abruptly, and Perrault appeared flummoxed about what to make of it, although it would have been the perfect opportunity for a comment on Satan's wiles. Near the end of the visitation, Perrault finally cursed the spirit, who then humbly told him that "You have lied. I'm not at all cursed. I hope for health through the death and passion of Jesus Christ."[44] This response left Perrault dumbfounded, silent while the spirit recited the Our Father.

While Perrault's sympathies were inappropriate if the spirit was a demon, certain aspects of Perrault's account sound like those found in ghost stories in early modern France and illustrate how the most committed Huguenot might accept, even if only tacitly, elements of traditional faith. Perrault's own explanation of the time when the spirit appeared to be two, distinct entities was that it behaved this way to convince them that it was the soul of a woman who had recently died in the house, that is, the daughter of the woman whom he had evicted from the house when he arrived in Mâcon. While it could be argued that Perrault was making this comparison to highlight the spirit's falseness—in the sense that the spirit thought that it could lead them into heretical belief in ghosts—Perrault's tone suggests this was not the case. Elsewhere Perrault uncritically recounts that the spirit of a recently dead woman had appeared to her relatives.[45] Like many apparitions of the dead, Perrault's spirit recited the Lord's Prayer, the Apostles' Creed, evening and morning prayers, and the Ten Commandments without any prompting or any mistakes.[46] When the spirit asked for assistance in resolving its will and deposition of property, it echoed a longstanding belief that ghosts were almost compelled to return if such business was unfinished. The spirit in Perrault's house also accepted the providentialism that legitimate spirits were supposed to show. At about the midpoint of the haunting, Perrault decided that he wanted to see what the spirit looked like, and the spirit said that it would only appear in physical form if it was God's will but that we should hope God "would soon deliver us from all temptation"—presumably the temptation to demand

---

[43]　Perrault never explains how a will could apply to a spirit, given that a will is a document produced by a living, corporeal human being and the spirit was presumed to be none of those things. For the will to have any possible legal validity, it would have to be proven that the spirit was a human, a conclusion Perrault rejects elsewhere and one certainly outside Reformed orthodoxy.

[44]　Ibid., 40, 42.

[45]　Ibid., 58.

[46]　Ibid., 29.

further proofs of the spirit's orthodoxy. Nothing more was said on this subject, and when the spirit finally took human form, it did so long after Perrault's request.[47] Each of these queries and tests was common in early modern ghost stories of either Catholic or Lutheran provenance but were less so among those with clearly Calvinist origins.[48]

More common and legitimate were the equations Perrault made between the spirit's visitation and magic, but in these statements, too, there was little distinctly Huguenot in his approach. His two primary explanations for the haunting were based on magic, not his flaws or God's providence. In one case Perrault stated that he was the victim of a magician named Caesar, who was angry that a royal decree had allowed the Huguenots to practice their religion openly in the town and region and who Perrault saw as acting in concert with some of Mâcon's town councilors.[49] Soon afterward Perrault hypothesized that the spirit appeared at the command of the woman who used to live in his house. When Perrault moved to Mâcon, she was evicted from the property where Perrault and his family planned to live and was heard cursing him in several places in the town.[50] Perrault avoided direct condemnation of these activities, merely noting them as causes of the pernicious apparition; as such, he adopted a perspective that would have offended few save those he accused. Surprisingly, though, he did not connect this woman and the apparition with witchcraft. In both Catholic and Huguenot territories throughout the region in the early seventeenth century, accusations of witchcraft were appearing to escalate, culminating in trials throughout Burgundy, the Franche-Comté, Savoy, and western Switzerland in the 1620s and 1630s. While Perrault may have been restrained, since the woman did cause him harm (*maleficia*), he also reflected an acceptance that some degree of magic, even damaging magic, could exist in a community without it being demon-ridden. In this case, Perrault showed himself a man of the Jura mountains.

Perrault's treatment of his own case was far from unique; he himself argued that his situation needed to be seen in the context of dozens of such events that were occurring throughout the region. Approximately one-third of the *Antidemon* is an attempt to explain why the spirit had beset him and his family "as there is nothing more common and ordinary, indeed more natural, especially

---

[47]   Ibid., 42.

[48]   Two recent books have provided extensive insight about continental perspectives towards ghosts in the sixteenth and seventeenth centuries: Timothy Chesters, *Ghost Stories in Late Renaissance France: Walking by Night* (Oxford: Oxford University Press, 2010); Miriam Rieger, *Der Teufel im Pfarrhaus: Gespenster, Geisterglaube und Besessenheit im Luthertum der Frühen Neuzeit* (Stuttgart: Franz Steiner, 2011).

[49]   Perrault, *L'Antidemon*, 62.

[50]   Ibid., 64–5.

when it involves extraordinary things, than to look for the cause."[51] His initial explanation was a litany of similar events. From Lyon to Paris and points between, Perrault describes "tragic and frightening" demonic apparitions of which he had personal reports. Mâcon itself was subject to multiple visitations.[52] He linked these visitations to a rise in witchcraft in the area and argued that the effects were enhanced within Mâcon because witchcraft trials were held in that city. While Perrault was rarely explicit about his sources, Lullier, the harassed goldsmith, made another appearance as an informant. In some cases, Perrault noted that the spirit left or was expelled, but in others he could not say what happened "despite conjurations and even some judicial procedures."[53] In his insistence on analyzing the spirit and his interest in juxtaposing it to other apparitions, Perrault strayed from the model of acceptance and prayer urged among Huguenot theologians and into the realm of curiosity and gossip that so frequently distinguishes these accounts, whatever their author's confession. As one among many victims of preternatural or supernatural activity, Perrault lost his unique and potentially dangerous status, but he also lost one cornerstone for claims to spiritual legitimacy and distinctiveness. God tested Catholics and Calvinists alike.

In the most thorough modern analysis of Perrault's *Antidemon of Mascon*, P.G. Maxwell-Stuart argues that there is "very little which is distinctively Protestant" in Perrault's account. What slight differences that might exist between a Catholic and a Calvinist exposition are tied to the "illustrative details" and the "remedies and safeguards the writer recommended."[54] In large part I agree. Here I have tried to expand this argument in more social and folkloric directions by exploring the importance of local networks in Perrault's depictions and the tensions an exemplary Protestant minister faced blending Catholic, Calvinist, and common traditions. They lead Perrault to stress the reality of his experience, not the more acceptable illusionary interpretation, and to call on practices and people from his community that might make his early teachers wrinkle their brows or shake their heads. In this sense the antidemon of Mâcon becomes something undemonic but too unpalatable to define.

---

[51]   Ibid., 55.

[52]   See Bost, *Histoire de l'Église protestante*, 207, for a list of supernatural events occurring in the Mâconnais at roughly the same time as Perrault's haunting.

[53]   Perrault, *L'Antidemon*, 57–62.

[54]   P.G. Maxwell-Stuart, "Rational Superstition," esp. 180. See his most recent analysis in *Poltergeists: A History of Violent Ghostly Phenomena* (Stroud: Amberley, 2012), ch. 6.

Chapter 10

The Constitution and Conditions of Everyday Magic in Late Medieval and Early Modern Catholic Europe[1]

Sarah Ferber

Recent history writing on the subject of "everyday magic" has seen it become something of a tugboat term that conveys a diverse range of methodological and cultural agendas, bringing into question the scope and foundations of earlier witchcraft scholarship. Scholars have proposed that stories of large western European witchcraft trials, especially those involving the use of torture and evidence about the witches' sabbat, are not representative of the range of shared magic and witchcraft beliefs or practices across Europe and the wider world.[2] In this view writing on everyday magic and "the magical universe," to use Stephen Wilson's expression, can provide a wider contextual map, for example, in relation to the history of witchcraft.[3] And histories of witchcraft and magic that remove them from debates about their legal status can serve to re-position their chronology and histories, de-emphasizing trials and drawing attention to continuities in practice from the medieval era to the present.[4] Wilson maintains that there are customs that "are universal in time and space" that "demand a

[1]   Thanks are due to Leigh Dale for her many invaluable suggestions for revision. This chapter is dedicated to the memory of my mother, Helen Ferber (1919–2013).

[2]   Valerie A. Kivelson, "Lethal Convictions: The Power of a Satanic Paradigm in Russian and European Witch Trials," *Magic, Ritual, and Witchcraft* 6:1 (2011): 34–61. Among the scholars whose work is discussed are Edward Bever, Marko Nenonen, Stephen Wilson, Wolfgang Behringer, and Erik Midelfort. It is important to note that the historians associated with these new directions do not share a uniform view of either the nature of the issues at hand nor the best way to address them.

[3]   Stephen Wilson, *The Magical Universe: Everyday Ritual and Magic in Pre-modern Europe* (London: Hambledon, 2000). See also Alexandra Walsham's review of Wilson's book: "Book Review of *The Magical Universe. Everyday ritual and magic in pre-modern Europe. By Stephen Wilson*," *Journal of Ecclesiastical History* 53:3 (2002): 594–5.

[4]   H.C. Erik Midelfort, "Witch Craze? Beyond the Legends of Panic," *Magic, Ritual, and Witchcraft* 6:1 (2011): 11–33, referring in particular to the work of Owen Davies, p. 29.

comparative approach to understand them."[5] Thus, the study of more widespread and less controversial magic can also lend support to the view that witchcraft trials were not always necessarily a sign of "panic."[6]

In 2011 Valerie Kivelson praised these revisionist developments but argued, specifically in relation to witchcraft history, that to speak of cultural differences as well as similarities remains important. She took issue with calls to reorient witchcraft studies to de-emphasize satanic and "exotic" aspects of witch beliefs because, she maintained, the presence of a backdrop of highly theorized demonology in western European witchcraft trials made this region distinctive.[7] She stated, "While the revisionist historians have convincingly shown that most magical practice throughout early modern Europe consisted of curses, spells, *maleficium*, or simple sorcery, unencumbered by the fanciful baggage of satanic lore, the very availability of a more theoretical framework of demonology sets the European case on an altogether different footing from that of much of the rest of the world."[8] Kivelson argued to retain culturally differentiated readings of magic and witchcraft as a necessary complement to a new trend of seeking evidence of the integrating features of magic and witchcraft across Europe and the wider world. This chapter seeks to build on Kivelson's argument, offering further observations about the history of magic and witchcraft in the Catholic jurisdictions of pre-modern western Europe. It will maintain that there is a risk in positioning everyday magic as a distinct set of behaviors and series of events historically parallel, but in essence unrelated to, stories of the witches' sabbat, for example, or to extreme instances of persecution of imagined witches. It would seem crucial to ask how events which were linked in time and place were also linked conceptually. To exclude a consideration of either change over time or of the unifying characteristics of a particular historical moment might, on the one hand, best be referred to by another name than historiography or, on the other, pose a fundamental challenge to the legitimacy of history writing.

The chapter will therefore hope to address these related questions: what context does knowledge of the everyday provide to help understand the occurrence of extreme events? To what extent can the nature of the everyday itself be understood by reference to extremes? In what specific ways were the everyday and the extreme linked in practice? The premise, here, is that the extreme and the everyday are categorically different only to the extent that some confluence of events sets off a re-categorization of the everyday into the category of the extreme—or the migration of the extreme into the everyday. New

---

5   Wilson, *The Magical Universe*, xxx.
6   Midelfort, "Witch Craze?"
7   Kivelson, "Lethal Convictions," 35.
8   Ibid., 38. See also Valerie Kivelson, *Desperate Magic: The Moral Economy of Witchcraft in Seventeenth-Century Russia* (Ithaca: Cornell University Press, 2013), 52–82.

readings of Reformation and witchcraft history in the late twentieth century emphasized the importance of seeing categories to which actions were assigned as always in a state of process. As Stuart Clark and Bengt Ankarloo emphasized in their introduction to the *Athlone History of Witchcraft*, magic "as a historical category ... is constantly created and recreated. It can therefore be understood only in relation to other categories which are also undergoing this process of continuous redefinition."[9] The quest for historians, then, can be to investigate ways the everyday and the extreme are linked in action and over time. How does one set of beliefs transform into a related set? Who sees a link where others have not? Who carries news or views from place to place or context to context in ways that permit new configurations of available narratives or the emergence of new narratives?

The focus here is on the origins and changeable status of everyday magic in Western Catholic Europe rather than on its function. From a methodological point of view the chapter seeks to emphasize the value of critical history to the understanding of historical change. Critical history seeks foremost to solve problems. Specifically it asks: what maintains the viability of systems of categorization? What factors influence decisions made about the category to which a practice—and therefore often a person—is assigned? How is the everyday constituted historically? What conditions underpin the choices available in everyday life? Given the realities of life, is the very idea of the everyday life itself more an aspirational statement than a reality? This chapter will suggest that what might be called everyday magic in Western Catholic history cannot fully be interpreted without reference to the authority that potentially construed all magical practices as a threat to appropriate worship. To the extent that the theology, material and ritual culture, and official structures of the church created the everyday, these structures represent the constant frame of reference through which the specific nature of not only everyday magic but indeed everyday life took place in Catholic Europe. Within this system, fear of demons was something of a constant presence, but it also needs to be understood alongside and in interaction with other sources of perceived adversity—notably pagans, Jews, and Muslims—enmity with whom was among the defining features of western Christendom. To explore these issues the chapter will draw on well-known examples from the historiography of early modern Europe, with qualifying comments, to make a case for understanding the history of everyday magic in Europe by reference to its geographical and historical specificity. This approach in turn is intended to draw attention to the need to avoid a

---

[9]   Stuart Clark and Bengt Ankarloo, "Introduction," in Karen Jolly, Catharina Raudvere, and Edward Peters, *Witchcraft and Magic in the Middle Ages*, vol. 3, in *The Athlone History of Witchcraft and Magic in Europe*, ed. Stuart Clark and Bengt Ankarloo (London: Athlone, 2002), x.

re-naturalizing of the very categories that can be used to speak of everyday magic. History scholars over the past 40 years have provided insistent reminders that the very categories of the "popular," "folk," "traditional," or of "superstition," "religion," and "magic," if they are to be used at all, need to be preserved in at least implied inverted commas.[10]

In the context of studies on everyday magic, it is important that uncritical use of such categories not impose modern assumptions about what they mean. For the Catholic West, the idea of everyday magic might suggest a magic belonging to the laity ("popular" or "folkloric"), which was "traditional"; possibly more pagan than Christian in origin; domestically and interpersonally oriented; instrumental and manipulative, rather than supplicatory; and in most cases unrelated to the work of the devil.[11] All these possible attributes were confounded, however, in late medieval and early modern Catholic Europe, and therefore, definitions are ultimately rhetorical, assertions made in the context of complex cultural alignments. Using the term "magic" can say more about what modern historians want to know or want to argue than about what people in the past thought they were doing: people in late medieval and early modern Catholic Europe who practiced what one might call everyday magic would not necessarily have seen themselves as performing a magical act at all. They might have seen an action as one of prayer or worship, or indeed as a straightforward act of healing, climate control, insecticide, or agriculture.[12] Likewise, whether or not a person

---

[10]   The logical and even nihilistic extension of this argument is that the meaning of no word is sufficiently stable to permit its use at any time. Therefore historians must, for their own narrative goals, permit each other a certain degree of latitude. As medievalist Judith M. Bennett has noted, "Historians are accustomed to using modern words to investigate past times, to assessing the changing meanings of words over time, and to weighing differences as well as similarities in their uses of such words.": "'Lesbian-Like' and the Social History of Lesbianisms," *Journal of the History of Sexuality* 9:1–2 (2000): 1–24, here 12. However, this pragmatism does not extend, as Bennett herself was aware, to ignoring the tentative nature of the history-writing enterprise. Rather, awareness of the importance of the historically contingent meanings of words and categories is an essential component of methodological self-awareness.

[11]   Karen Jolly, "Definitions of Magic," in Jolly, Raudvere, and Peters, *Witchcraft and Magic in the Middle Ages*, 8. See also R.W. Scribner, "Cosmic Order and Daily Life: Sacred and Secular in Pre-industrial German Society," in *Popular Culture and Popular Movements in Reformation Germany* (London and Ronceverte: Hambledon Press, 1987), 12.

[12]   R.W. Scribner cites one example of this diversity of applications from the work of fifteenth-century theologian Joannis de Turrecremata, who saw the effects of holy water as divisible into three categories. As Scribner described them, "Four of its effects were spiritual or moral: recalling the heart from earthly things, remitting venial sin, as preparation for prayer or for doing good works. Two were psychic or psychological: purifying the mind from fantasies and driving out impure spirits. The last four were a matter of direct physical efficacy: removing infertility in humans and animals, encouraging fertility in all earthly

was causing good or harm by casting a spell might well have depended upon who acted, who was meant to be affected, and how they were affected. Miracles, magic, and even witchcraft would not necessarily have been seen as categorically different, despite theological arguments to the contrary. Even considering magic at its seemingly most crude, humdrum, or uncontroversial, how is it possible to write about power-seeking behaviors without writing about the other systems of power within which people practiced them? This comes down to a question of what kind of history is worth writing.

Readers familiar with recent debates in early modern history writing will see the goal of "re-problematizing categories" as entailing the use of two historiographical trends of the late twentieth century: the so-called narrative turn and the linguistic turn. These methodological developments arose principally in two emerging areas of study: social history of the Reformation and the related but also distinctive and emerging history of magic and witchcraft. A wider backdrop was the emergence of cultural studies, especially the study of so-called popular culture, reflected in an interest in oral history and folk history, to some extent grounded in a Marxian interest in "ordinary people."[13] An extensive historiography from the 1970s to the present challenged traditional assumptions about the given-ness of categories even as seemingly basic as religion and magic. This body of work provided a conceptual frame for interpreting the vernacular magic common to much of Christian Western Europe. But for many historians writing in this vein, the category of "the popular" itself was open to question: what to make of the fact that bishops exorcized plagues of locusts, for example? Similarly, terms such as "traditional" magic or "superstition" came under scrutiny, as historians made clear that, if categories which were used at the time were to be uncritically reproduced in modern research, it might seem as though the modern authors were either ignorant of their subject's context or sympathized with the views of those who created the categories.[14] The continuum between miracle, magic, and witchcraft was not one of simple overlap between fixed categories. Because there existed in Catholic Europe a system for ranking behaviors according to whether or not they reflected an appropriate form of

---

things, repelling pestilence and protecting against sickness." From the pen of a theologian whose task it was to assert differences between religion and magic where none might be apparent, the example provides paradoxical proof of the subtle and even arbitrary nature of such a distinction: Scribner, "Cosmic Order and Daily Life," 11.

[13]   See Carlos M. Eire, "The Concept of Popular Religion," in *Local Religion in Colonial Mexico*, ed. Martin Austin Nesvig (Albuquerque: University of New Mexico Press, 2006), 1–36.

[14]   Stuart Clark, "One-Tier History," *Magic, Ritual, and Witchcraft* 5:1 (2010): 84–90; Stuart Clark, "Magic and Witchcraft," in *Finding Europe: Discourses on Margins, Communities, Images ca. 13th–18th Centuries*, eds A. Molho, D.R. Curto and N. Koniordos (New York: Berghahn Books, 2007), 115–30.

worship, using terms such as religion, magic, and witchcraft is as much about describing a hierarchy as it is about describing what some people thought they were doing. The idea of a continuum speaks of the countless cultural scenarios in which activities that, to modern readers, might seem similar were historically categorized as one thing or the other as a result of perception, analysis, and active intervention. The contents of the categories matter less than the value system that determined which categories of activity belonged where in a moral or legal hierarchy.

Determining where lines were to be drawn was a profound preoccupation of the late medieval and early modern era. A process of accelerated demonization occurred in at least two ways: at the level of religious reform in campaigns against so-called superstition, in which devil-worship was imputed either actively or implicitly to the practice of magic; and in witchcraft trials, as the course of individual trials unfolded and allegations shifted from interpersonal infractions among neighbors to confessions of trafficking with demons. Superstition was a theological category that referred to an assertion of the need to separate Christian from non-Christian. It was not simply descriptive: the very use of the term indicated that there was a problem to be addressed.[15]

In this light, to explore the *constitution* and *conditions* of everyday magic points to two lines of investigation. First, it becomes possible to provide evidence that the *constitution* of everyday magic for pre-modern Catholics was far from everyday in itself. By the later Middle Ages, a large proportion of the magic practiced within the jurisdictions of the Roman Church contained elements of Catholic lore, founded on dramatic stories about suffering, divine power, and struggle against the enemies of Christendom and demonic forces.[16] These were stories of Christ, of victories over the devil, of saints in life and their miracles after death: they are what endowed the liturgical and paraliturgical activities of Catholic Europe with power. Magic might seem to have been everyday, because of the often mundane ends to which it was put or its occurrence within an accepted ritual cycle. But it could not be separated from the dramatic and compelling narratives that underpinned the power attributed to everyday rituals and objects of Catholic magic. Efficacious uses of blessed objects or rituals worked for reasons that were fundamentally not physical but moral; the suffering and altruism of Christ and the holiness of Mary and the saints provided not only models of goodness and selflessness, but the moral force that drove the machinery of miracles. The source of material outcomes was thus both

---

[15]   Categories were not infinitely elastic, of course. While there could be a variety of responses to the same kind of activities from within the church itself, some charges were also more likely to stick: misuse of the Eucharist, for example, was indefensible.

[16]   Edward Bever, "Popular Witch Beliefs and Magical Practices," in *The Oxford Handbook of Witchcraft in Early Modern Europe and Colonial America*, ed. Brian P. Levack (New York: Oxford University Press, 2013), 50–68, here 52.

numinous and moral, found in the stories of sacred persons who were entitled in differing degrees to worship or veneration and who provided for their part either direct aid or mediation with the divinity. In recognizing the suffering of mortals, saints entreated God on their behalf to grant mercy. Everyday magic might be mundane in its application, but theologically the power to heal might derive from painful sacrifice, turned to good account. As one German prayer encapsulated it: "Oxen, bear the yoke and be patient, as Christ was patient."[17]

Theologically, prayers or other help-seeking devotions could not compel any action by the saint or God. While some formal advice to supplicants did suggest that the effects of their actions could be automatic, such advice must be understood as being more likely conveyed in the process of pastoral care than as an expression of theological precision. What might have been experienced as a system of exchange—beginning with an offering of a prayer or good work—was not technically equivalent to service provision, though it amounted to one culturally and even found a measure of theological support.[18] This system operated through the sacraments and sacramentals: a moral power was the active ingredient of the objects, times, and places associated with holy people. When a priest blessed holy water on a saint's day, for example, he conveyed to normal water through his own sacralized status some of the goodness of that saint. The invisibility of such moral exchange, however, opened the way for potentially endless inflation of the expectations of what could be achieved through devotional/magical acts, which could ultimately invite ridicule or accusations of diabolism. At this point, the conditions within which people were able to practice magic licitly comes into focus, since everyday magic can be understood as always susceptible to retranslation into a category of either harmful magic and/or devil worship. Moreover, because magical activities came under implicit church surveillance, they cannot be treated as a finite category; rather, religiosity, ritual, and magic need to be understood as subject to continuing and often competing processes of categorization. Not all churchmen agreed on the detail, but a capacity to assign magical behaviors to categories was one of the sources of church authority. Therefore in considering the *conditions* of everyday magic, the second line of investigation here, the chapter will evaluate the authoritative decision-making which classified some activities as illicit magic or witchcraft—distinct from acts of appropriate worship—and thus as pernicious.

The three following sections present, first, a more detailed outline of the constitutive role of narrative in the belief system that sustained most of the "magic of the medieval church," to use Keith Thomas's evocative and contentious

---

[17]   Scribner, "Cosmic Order and Daily Life," 11.

[18]   Scribner discusses this question in relation to the writings of the fifteenth-century reforming priest Geiler of Kaisersberg: Ibid., 12.

term.[19] Next, the chapter will consider the system of official church practice that provided the conditions through which activities were determined licit or illicit. Finally, the chapter will consider demonic possession and witchcraft cases from France in the early modern period. We shall see that early modern cases that could be seen as exotic and extreme could, just like earlier dramas within Catholic history, find their way into everyday magic. Building on the arguments of the first two sections, the history of these cases shows that everyday magic could be linked in a range of ways to some of the most extreme manifestations of Catholic Reform religiosity.

Stories of Power

Even given the wide diversity of local traditions and Christian practice, by the late medieval era the Roman Church was sufficiently entrenched throughout western Europe that its healing and blessing activities, as well as its work in pursuit of salvation, had become part of the fabric of everyday life for all those baptized into the faith. The church's magic was poised at the juncture of two worlds. Its goal was always to orient believers to honor God and to think about the need for salvation, but its means were often material, aiding people in their own, their family's, or their community's worldly, emotional, and spiritual lives. The "church magic" of Roman Catholicism built on the marginalized magic of Celtic, Slavic, Germanic, and Scandinavian traditions of western, southern, and northern Europe. It also bore with it elements of Greco-Roman cults and of Jewish traditions, such as the rite of exorcism.[20] Western Christianity thus absorbed at the same time as it opposed the magic of its rivals. The resultant syncretist "system of the sacred," as David Gentilcore described it, permitted people at all social levels to pursue remedies or happier lives through their own devices and through interactions with a wide range of service providers. These providers included priests, witches, other healers, and diviners, whose resources were often used indiscriminately or in sequence.[21]

When European pagan rulers became Christian, whether by force or persuasion, the symbolic moment of their acceptance was the rite of baptism, a sacrament which the "vast majority" of people in medieval western Christendom

---

[19]   Keith Thomas, *Religion and the Decline of Magic* (Harmondsworth: Penguin, 1973), 27–57.

[20]   Jolly, "Definitions of Magic," 4; Bever, "Popular Witch Beliefs and Magical Practices."

[21]   David Gentilcore, *From Bishop to Witch: The System of the Sacred in Early Modern Terra d'Otranto* (New York: Manchester University Press, distributed by St. Martin's, 1992); Mary R. O'Neil, "'Sacerdote ovvero strione': Ecclesiastical & Superstitious Remedies in 16th Century Italy," in *Understanding Popular Culture*, ed. Steven Kaplan (Berlin: Mouton, 1984), 53–83.

underwent.[22] The rites of baptism, confirmation, and extreme unction marked the human fundamentals of birth, maturation, and death. The life of every Christian began with a ritual based on a story. The story was of the devil's quest to deter Christians from living in a godly way and the ritual to fight this battle was the rite of pre-baptismal exorcism that preceded the sacrament of baptism.[23] Every child's spiritual journey began with a godparent's renunciation of the power of the devil. In this simple but fundamental way, a reminder of the presence of the devil as God's enemy was entrenched in the foundational human rite of passage. The story of the devil's pursuit of Christian souls and his ritual expulsion fits with the idea that powerful narratives underpinned cyclical and mundane events.

The Western Catholic calendar and the historical stories that gave power to rituals and objects can help us to see the narrative constitution of everyday magic. The Christian calendar was in essence a story of holy people whose lives and afterlives endowed the astronomical year with a moral meaning. Day, night, and season were framed by a calendar of hotspots of which some elements were common, such as Christ-centered feasts, while others were more locally defined, usually by reference to saints' days.[24] Underlying the annual liturgical cycle were compelling narratives of lives combining the ordinary and extraordinary. The Easter story of Christ's death and resurrection was one of injustice, bravery, sacrifice, violence, and martyrdom, followed by a crowning miracle in the resurrection. The violence of this story was made visible in images, statues, and other artifacts, including the many claimed pieces of the True Cross which circulated throughout western Europe.

Saints, like Christ, were not only powerful entities; they were both an embodiment and source of two kinds of stories: the stories of what made them saints (such as the violent martyrdoms of St. Sebastian or St. Stephen at the hands of the Romans) and the accumulated stories of their deeds. The idea of a religion perennially on a war footing—whether against pagans, Muslims, Jews, heretics, and eventually witches—touched elite and non-elite medieval Europeans alike from the High Middle Ages well into the early modern era. Many saints, whose lives provided the compass points for the magical life of western Europeans, were portrayed as victims of other people and, importantly, as victims of pagans, while Christ himself was portrayed as the victim of

---

[22]   Robert Bartlett, *The Making of Europe: Conquest, Colonization, and Cultural Change, 950–1350* (Princeton: Princeton University Press, 1994), 251.

[23]   See John Bossy, *Christianity in the West, 1400–1700* (Oxford: Oxford University Press, 1985), 14–19; Bodo Nischan, "The exorcism controversy and baptism in the late reformation," *Sixteenth Century Journal*, 18:1 (1987): 31–51.

[24]   R.W. Scribner, "Ritual and Popular Religion in Catholic Germany at the time of the Reformation," in *Popular Culture and Popular Movements*, 17–48, here 19–22 (London and Ronceverte: Hambledon Press, 1987).

the Jews. The everyday was undergirded by an association with miraculous and often violent tales of Roman Catholic history; both the miraculous and the demonic in Catholic history pulsated beneath everyday life, providing the conditions of both unity and division. An overarching narrative of religious enmity had the capacity to sustain the idea of Christendom under siege and gave value to the lives of those who fought for it. Stories of what saints had done in the service of both the church when alive and, when dead, for communities who worshiped them, were a dynamic source of proof of contact with the supernatural.

Stories in turn penetrated everyday life in ritual charms, times, and blessings, as well as in the ongoing presence of household objects that could be sacralized, such as candles. All of these might be used in daily life, but they were rendered holy through a priestly blessing on another, auspicious day. Two examples described for Germany are the blessing of water on the feast days of St. Blasius and of St. Stephen, each a church martyr; on each of their days water was blessed for feeding to cattle and other livestock to provide year-round protection from illness or theft.[25] The story of Jesus's life was also re-told repeatedly across the year in the miracle of the Mass, which re-enacted both the sacrifice and the triumph of its subject over death itself. The sacrament generated one of the most holy objects of the late medieval era, the Host, portable and reproducible, a ubiquitous relic, especially at that time. The Host was widely regarded as a powerful force for fighting demons, for protecting communities and their resources and for healing.[26] The everyday was suffused with the miraculous. In the words of Benedicta Ward, miracles represented "the ordinary life of heaven made manifest in earthly affairs, chinks in the barriers between heaven and earth, a situation in which not to have miracles was a cause of surprise, terror and dismay."[27]

Positive saintly magic was part of what John Bossy referred to as the social miracle: saints' stories and the miracles they dispensed bound communities together in crucial ways.[28] In a famous archive of local religion collected under the auspices of Philip II of Spain and analyzed by William Christian Jr., people in Spain were invited to provide, among other accounts of their religious lives, the

---

[25]   R.W. Scribner, "Ritual and Popular Belief at the time of the Reformation," 17–48, here 33. Scribner provided in diagram form an overview of the categories. Another diagram showed how Protestant reform winnowed out clerically approved magic to leave in theory at least a more spiritualized religion that required fewer visible demonstrations of divine power. For a critique of aspects of his analysis, see Eire, "The Concept of Popular Religion," 18–21.

[26]   Charles Zika, "Hosts, Processions and Pilgrimages," 36.

[27]   Quoted in Gentilcore, *From Bishop to Witch*, 168. See also Scribner, "Cosmic Order and Daily Life," 12.

[28]   Bossy, *Christianity in the West*, 57.

stories of their towns' local memories of saintly intervention.[29] These stories were part of the lifeblood of the everyday. The power of such narratives substantially created the web that held together everyday life. Seen in this light, the imaginary is not readily teased out from the practical and material world, for the everyday is always situated at the intersection of a series of historical processes which, on the one hand, can be taking place in a moment or the space of a few years and, at the same time, reflect changes and beliefs which have evolved across centuries. A narrative community was shaped through imagining the lives and stories which endowed magical rites and objects with power. To reduce magic to its pragmatic, material, or instrumental uses might deny the importance of the imagination and underestimate the effects of stories on individuals and communities.

Systems of Authority

Broadly conceived, for Catholicism, magic was worthy of condemnation when a person culpably mistook themselves, their rituals, or the non-approved forces they conjured as the source of their actions' effects. Such a mistaken belief is one of the key nodes linking culpable magic and criminal witchcraft.[30] Theologically, all power derived from God/Christ and all worship was due to Him or, in limited ways, to Mary or the saints as intermediaries. To believe that anything other than God could bring about magical ends was implicitly to turn away from Him. Thus, the practice of magic, whether or not others were harmed, could be a serious offence.[31] Active or tacit clerical approval of healing or ritual activities was a first step to legitimacy. Importantly, however, Kieckhefer underscores a fundamental difficulty in "distinguishing in the concrete between orthopraxy on the one hand, magic and superstition on the other." He argues that, "What remains important is that writers on magic insisted there was a distinction, and crossing the line was a serious offence, a flirtation with demonic aid, even if, for an outsider, the

---

[29]   William A. Christian Jr, *Local Religion in Sixteenth-Century Spain* (Princeton: Princeton University Press, 1981).

[30]   P.G. Maxwell-Stuart, ed. and trans., *The Occult in Early Modern Europe* (London: Palgrave Macmillan, 1999), 115. Even distinctions such as that between the efficacy of the seven sacraments (which theologically was automatic) and that of the paraliturgical sacramentals (which in essence called for a one-off divine action) could be blurred, however. For example, Geiler of Kaisersberg, the reforming theologian, maintained that as long as devotion was in the heart of the practitioner the action of a sacramental might be automatically efficacious even without a process of saintly mediation or divine deliberation: Scribner, "Cosmic Order and Daily Life," 12.

[31]   Richard Kieckhefer, "Magic and its Hazards in the Late Medieval West," in Levack, *The Oxford Handbook*, 13–31.

line seems invisible."[32] Magic was an activity that was explicitly to be separated from religion through a process of theological analysis and classification. The church was an international corporation that made decisions about which activities were to be classified as magical. All magic practiced within Roman Catholic territories—whether "church magic" or magic with non-Christian origins in either literate or non-literate sub-cultures—was subject to church rules simply by virtue of its location within a Catholic territory. The church was composed of extensive networks and hierarchies with changing priorities, however: approval by one person at one time did not imply ongoing approval. A variety of changeable conditions affected the process of categorization. As Karen Jolly outlines,

> [t]he symbiotic relationship between the common tradition and the elite definitions of magic, each shaping the other, creates a large grey area of popular practices in Christian Europe that are not clearly magic or miracle, but lie on a spectrum in between. The popular practices lying across the middle of this spectrum ... are constantly adapted to changing circumstances, including the ecclesiastical and intellectual winds of change. In turn, the intellectual changes in the magic-religion paradigm, evolving out of new assumptions about knowledge and nature, then seek to recategorize existing practices toward one end of the spectrum or the other.[33]

Viewed through the lens of official oversight, activities experienced or practiced as everyday could become redefined as unacceptable or indeed reabsorbed back into everyday life. The system was not total and there were often substantial differences of opinion within the church, frequently along hierarchical lines from local priest to bishop or above. The institutions for scrutiny, notably the official hierarchy and, increasingly from the later Middle Ages, the inquisitions, were media through which definitions were imposed. The category of the everyday was thus not so much a set of activities as it was a status at once aspirational for the practitioner and conditional for the church. Only once it is established that distinctions between categories are not fixed, but are subject to a changeable system that serves to grant legitimacy, does it become possible to historicize magic seriously.

From the early Middle Ages, as European pagans learned how to be Christian, an ongoing process of separation of pagan worship, which was in the official Christian view at least implicitly diabolical, from correct worship redefined the

---

32   Kieckhefer, "Magic and its Hazards in the Late Medieval West," 15.
33   Jolly, "Definitions of Magic," 6–7.

spiritual jurisdiction of Europe.[34] The church's continued commitment to using either sacred objects such as relics or material objects blessed by priests created an inevitable tension, as the portability and ready availability of blessed objects also spread the risk that use of these objects might be seen as illicit.[35] Similarly, apparent miracles of holy people, such as visions, levitation, or transports beyond natural capacities all opened up potential for debate and disagreement.[36] A few well-known examples can serve as illustrations.

A 1984 study by Mary C. O'Neil focused on the difficulties lesser clergy faced in relation to healing and divinatory uses of religious paraphernalia as they sought to balance the expectations of parishioners and deflect the suspicions of inquisitors.[37] Similarly, in the often-cited history of the Italian fertility cult of the *benandanti*, priests actively participated in sanctifying the community beliefs by blessing the cauls of newborns believed to be chosen to take part in nocturnal fights for crops. In these examples as in many other cases, Catholic inquisitors moved to enforce what they held to be the line between "superstitious" and licit behaviors.[38] Similarly, although not directly involving magic, an account of the rise and fall of a bleeding Host shrine in the bishopric of Passau in fifteenth-century Germany shows the official machinery of categorization at work: a stamp of authenticity rapidly provided by authorities at a lower level in the hierarchy was undone because of intervention from higher up.[39] A peasant and his sons had purportedly found "bleeding Hosts" in a container hidden in thorn bushes,

---

[34] Jean-Claude Schmitt, *The Holy Greyhound: Guinefort, Healer of Children since the Thirteenth Century*, trans. Martin Thom (Cambridge: Cambridge University Press, 1983), 22. See also Valerie I.J. Flint, *The Rise of Magic in Early Medieval Europe* (Princeton: Princeton University Press, 1994).

[35] Lyndal Roper, "Exorcism and the Theology of the Body," in *Oedipus and the Devil: Witchcraft, Sexuality and Religion in Early Modern Europe* (New York: Routledge 1994), 171–98, here 172–4.

[36] See the valuable discussion of the authentication of claims about holy women in Anne Jacobson Schutte, ed. and trans., *Cecilia Ferrazzi: Autobiography of an Aspiring Saint* (Chicago: University of Chicago Press, 1996), 14–16.

[37] O'Neil, "'Sacerdote ovvero strione,'" 53–83.

[38] Carlo Ginzburg, *The Night Battles: Witchcraft and Agrarian Cults in the Sixteenth and Seventeenth Centuries*, trans. John and Anne Tedeschi (London: Routledge & Kegan Paul, 1983), 9. Here I use "superstitious" in the sense understood in late medieval and early modern Europe, as an inappropriate form of worship linked theologically to worship of the devil.

[39] The report of the case by Dominican Heinrich Institoris, author of the *Malleus maleficarum*, does not mention the use of the shrine for healing or other secondary purposes beyond worship. However, the practice of gazing at the exposed Host, common for the region and time, might mean that such a use could not be excluded from the activities that had taken place at the incipient shrine. Charles Zika, "Hosts, Processions and Pilgrimages: Controlling the Sacred in Fifteenth-Century Germany," *Past and Present* 118 (1988): 25–64, here 31–3. The article is reproduced in Charles Zika, *Exorcising our Demons: Magic, Witchcraft and*

and a local provost had certified that the Hosts were "the sacrament of the Eucharist," approving the establishment of a makeshift shrine suitable for local people to visit.[40] Worship of the Hosts at the local church was not approved; the bishop of Passau went further, moving to quell the incipient cult entirely. He had the shrine dismantled and told his priests to preach that local people had been worshipping not Christ but "the devil, as well as thorn-bushes, trees and the rust of the containers."[41] In a short space of time, therefore, an object of devotion was legitimated, then re-paganized, hereticized, and demonized. The status of the everyday was thus substantially contingent on the good grace of authorities.[42]

The trial of Joan of Arc perhaps most conspicuously exposes the capacity of different church jurisdictions, legally convoked, to both condemn and defend the same actions. In 1431, Joan was "denounced and declared ... a witch, enchantress, false prophet, a caller-up of evil spirits, ... superstitious, implicated in and given to magic arts, thinking evil in our Catholic faith, schismatic in the article Unam Sanctam, etc. ..."[43] In 1456, under the authority of Pope Calixtus III the trial outcome was nullified and its findings overturned. In each of these cases, the line drawn between licit and illicit magic—between everyday and diabolical—is as much a political and hierarchical one as it is an absolute distinction.

The records of the Venetian Inquisition provided historian Sally Scully with two important case studies that permit us to see everyday magic and witchcraft in the working life of two women, half-sisters, in seventeenth-century Venice. Scully's goal was to incorporate the history of witchcraft into the history of labor, arguing that "the witch's hat was one of many, taken off and put on at will, signifying a vocational choice rather than a permanently assumed role."[44] The sisters' magic was not everyday church magic: in these cases it was non-Christian magic adjudged diabolical under the watchful eye of the Inquisition. One of the women, Marietta Battaglia, was put on trial for both diabolical magic

---

Visual Culture in Early Modern Europe (Leiden: Brill, 2003), 155–96. Thanks are due to Charles Zika for clarification of the stated reasons for the shrine's closure.

[40]   Zika, "Hosts, Processions and Pilgrimages," 27.

[41]   Ibid., 29.

[42]   For a new set of approaches to this question, see Eire, "The Concept of Popular Religion," 23–7. The process was not all one-way, however. Authorities were not uniform in their desire to condemn incipient cults. A more usual trend was the higher up the hierarchy, the more likely a cult was to be shut down. Moreover, categorizations of folk-versus-official; popular-versus-official or popular-versus-learned; or lay-versus-clerical are too procrustean to tell the complex story of the ways in which magic, devil worship, or the worship of God have come historically to be separated.

[43]   http://www.fordham.edu/Halsall/basis/joanofarc-trial.asp (accessed 2 January 2014).

[44]   Sally Scully, "Marriage or a career? Witchcraft as an Alternative in Seventeenth-Century Venice," *Journal of Social History* 28:4 (1995): 857–76, here 857.

and prostitution, activities which Scully argues were simply parts of a quest for income. Scully refers to Battaglia's "[m]ultiple and undifferentiated magical and sexual activity [which were] ... supplemented by cooking."[45] Found guilty, Battaglia's sentences were commuted if she agreed to "cook for half-wages at the Arsenale."[46] The case of Battaglia's sister, Laura Malipiero, is equally notable. Her magical activity included both paid magical assistance to others and the more unusual activity of maintaining premises where books of magic were hand copied. Malipiero was tried four times and punishments included a prison term, but she died a free woman of some means. Scully's study shows that the Inquisition repeatedly exercised its authority to redefine as harmful magic what Sully argues was "just a job," and that for both women, "[w]itchcraft practice was an extension of their composite life-pattern."[47] Scully convincingly argues the case for the absorption of the sisters' magic into their everyday lives. But the stories of these two women also show that their capacity to practice such everyday magic was nonetheless conditional on the repeated involvement of the Venetian Inquisition defining and controlling their activities.[48]

The Domestication of the Demonic

In the late medieval and early modern era, extreme events arising from emerging practices were reinvested into mainstream religious and community life alongside the historic stories of holy people. Just as the histories of martyrs were suitable bases for cults to grow around particular saints, saintly suffering served as a model for the experience and interpretation of demonic possession. Increasingly from the late medieval period, the discourse of saintly struggle against the power of demons became in many cases assimilated to the experience of demonic possession, associated with bewitchment. Thus even notorious French cases of demonic possession, which are sometimes credited with being among the most exotic and unrepresentative Western stories of witches and demons, provide evidence that can contribute to meaningful discussion of the everyday. Such cases became suitable for domestication not in spite of but because of the extreme behaviors to which they bore witness.

Marthe Brossier was a lay demoniac whose performances under the charge of exorcists led her to become a weapon in the war of zealot Catholics against

---

[45]   Ibid., 860. Scully's article further underscores the somewhat random nature of magic and witchcraft charges by referring to the doubtful motives of disgruntled or former lovers who accused the sisters of witchcraft, among a range of other offences.

[46]   Ibid., 860.

[47]   Ibid., 858.

[48]   Scully's argument also anticipates critiques of the notion of a witch stereotype encountered in recent discussions of everyday magic.

the Huguenots around the time of the Edict of Nantes (1598). Like many holy women and demoniacs, Brossier was held to be capable of reporting back from nether worlds or distant places accessible only in states of ecstasy, possession, or trance. In towns where she performed, people perceived Brossier as capable of providing personal information of the kind they might otherwise have sought from a local diviner-healer. In Romorantin, for example, before Brossier embarked on a lengthy journey to Paris to work with her exorcists against the Huguenots, people who attended her exorcisms asked her (or, technically, her demon) to tell them if parents were in heaven or purgatory, if enemies would be damned when they died, and if husbands working away would come home safe.[49] It is not clear how much her handlers envisaged the functions of a seer as selling points or how much they were merely tolerated, but they did fuel the ridicule which many *politiques*—Catholic supporters of Henri IV's attempts at peace—chose to bestow on the possessed. The *politiques* saw women like Brossier as easy targets through which to foil their theological opponents, by accusing them of being no better than magicians or travelling showmen. *Politique* critics made explicit comparison of public exorcism to the shows put on by jugglers and performing bears; this reinforcement of the similarity of demonic possession with the frivolous, bestial, and non-magical—one might say the truly everyday—was an attempt to expose bad faith. In a religious tradition reliant on manifestations of the spirit world or the supernatural in the material world, there was an inevitable tension at work that rendered categorical distinctions inseparable from expressions of authority. Again, such assertions need to be understood as acts of political differentiation, not incontrovertible statements of absolutes. But Brossier's story reinforces, foremost, the view that the exotic could not only be channeled into the everyday, but was most suitable for such use. Her journey across France appears to have fitted into everyday life for villagers whose existence was likely to have been regularly punctuated by itinerant preachers, flagellants, or pilgrims. The community service that Brossier rendered by reporting on the status of absent locals belonged on the spectrum of recognized media for communications between the living and the dead. She was seen as a "communal resource," if a transient one.

Brossier was by no means the only woman in early modern France who negotiated such apparent extremes. Another lay possessed woman, the Norman mystic Marie des Vallées, took on the identity of a lifelong demoniac, integrating in her lifestyle the magical, miraculous, and mundane. Holy possession as a form of living martyrdom was for her a way of life, and for many decades there were no serious attempts to exorcise her demons. Her greatest admirer, Father Jean Eudes, was canonized, although, it has to be said, in spite of, not

---

[49]   Sarah Ferber, *Demonic Possession and Exorcism in Early Modern France* (London: Routledge, 2004), 42.

because of, his admiration for Vallées. Even though she made use of expressions such as "deification"[50] to describe her own condition and was referred to in some circles as a "female messiah,"[51] for locals in Coutances Marie des Vallées was the housekeeper of two priests who could be seen publicly absorbed in her devotions—on one occasion to the merriment of local children.[52] Vallées claimed to have made a journey into hell to redeem witches of their sin, but as a jeering commentator observed, she had in fact been seen about town, leading a normal life.[53] What looked like the everyday life of a slightly eccentric local woman masked an intensive spiritual journey, suggesting yet another way in which the everyday and the extreme could be linked through narrative: in this case in the (auto-)hagiography of a mystic.

Another lay demoniac, Elisabeth de Ranfaing, integrated everyday and extraordinary, orthodox and dangerous. Hers was the story of a devout woman who claimed to have fallen victim to the most foul witchcraft in the form of a very everyday magical object: a piece of salt pork proffered by a suitor. The alleged witch, a local doctor names Charles, was executed on the basis of her accusations. Not long after, Ranfaing founded a religious order in Nancy where a devotional cult, known as the "médaillistes" also grew up around her experience of bewitchment and possession.[54] Several local Jesuits actively nurtured the cult, which derived its frisson from the dramatic story of her bewitchment and ultimate exorcism. Ranfaing's supporters sold little medals, which she herself blessed, thereby relaying and multiplying the holiness she was believed to have attained through her suffering. "Médaillistes" used the holy images depicting Jesus and Mary and bearing names of God written in Hebrew, marked with a cross above and below along with (in the words of a senior Jesuit opponent) "other gimmicks, to cast aside scruples, heal headaches, excite passion and appease quarrels, producing most often very pernicious effects."[55] Notable here is that such everyday forms of magic, involving health care, love, and even

---

50   Bibliothèque Mazarine, Paris, MS 3177, 243.

51   Bibliothèque Nationale, Paris, [hereafter cited as BN] Mss Fds fr., 14563, fol. 4v.

52   Ferber, *Demonic Possession*, 128, 132. "Housekeeper" is sometimes read as a code word for concubine, but nothing about this case suggests that Vallées was one.

53   Ibid., 130.

54   Sarah Ferber, "Cultivating Charisma: Elisabeth de Ranfaing and the *Médailliste* Cult in Seventeenth-Century Lorraine," in *Rituals, Images and Words: Varieties of Cultural Expression in Late Medieval and Early Modern Europe*, eds F.W. Kent and Charles Zika (Turnhout: Brepols, 2005), 55–84.

55   "Pièces Justificatives," 12, [BN Ms fds fs 494, fols. 824–25], in Etienne Delcambre and Jean Lhermitte, *Un cas énigmatique de possession diabolique en Lorraine au XVIIe siècle. Elisabeth de Ranfaing, l'énérgumène de Nancy, fondatrice de l'ordre du Refuge. Etude historique et psychomédicale* (Nancy: Société d'archéologie lorraine; Impr. Société d'impressions typographiques, 1956), 135.

peacemaking magic are depicted as intolerable. The Ranfaing case shows how one group of priests was able to domesticate a cult based on very extreme events and to redirect those energies into an everyday project of magic and lay devotion. As with the stories of saints, local people would have known the back-story of this drama, and many appeared at ease with and inspired by the version of Ranfaing's supporters. Diabolical links did not therefore solely undermine or challenge the value of church magic in everyday life. The process could work two ways: common magic could be demonized and turned into witchcraft accusations, but as this example shows, equally, stories of the demonic and witchcraft could be re-absorbed back into a porous everyday culture as a propellant of charisma. A continuum between the everyday and the extreme is evident, therefore, even in such seemingly Epicurean events as cases of demonic possession and witchcraft.

Conclusion

This chapter has argued that the "everyday-ness" of magic in pre-modern Catholic Europe was established not by reference to those who practiced it or the nature of the practice; it was perpetuated and given value through the agency of the Roman Church. When campaigns against superstition took place, what mattered was the heading under which a behavior was categorized. For a belief or behavior to remain in the category of the everyday was conditional: an impression of continuity can belie a state of contingency. The first section of the chapter argued that everyday magic in Catholic tradition was constituted in and powered by the narratives of Catholic history, both local and trans-local. The second argued that the classification of actions as licit magic, illicit magic, witchcraft, or devil worship was a process reflecting the historical and institutional conditions surrounding perceptions of a wide range of activities. The characteristics of what lay in the category of the everyday, therefore, were subject to their status in the context of a system of authority that both permitted and ruled out a large variety of activities. We have considered here processes of change in the context of an understanding of history that sees not so much the categories, but the power to impose categories. Rather than seeing the everyday as an impermeable or timeless set of practices, the everyday, the new, and the ideologically charged need not be seen as categorically different. western European magic was defined in relation to the religion of which it was a necessary part but with which it coexisted in tension. Scrutiny and categorization continually defined the magic of Catholic Europe in its inevitably tense relations with its own pagan history and the productive tensions inherent in its sacramental theology.

Western European church magic itself was, plausibly, built on the violence, vitality, and drama of the narratives and magic of the Christian tradition, interwoven with the traditions that were in existence as Europe became Christian. A complex dynamic of mainstream and marginal, legitimacy and pretence, falsehood and authenticity characterized a Roman Christian culture in the late medieval and early modern eras. The capacity to remain in the everyday in turn relied on staying within the bounds of a series of binaries, on the other side of which potentially lay severe punishment. Legitimate activity was always defined against the possibility of illegitimate activity, just as the heroism of Christian figures positioned them historically against a diverse range of adversaries. The constitutive elements of pre-modern magic, and the conditions under which it was practiced, reveal no categorical difference between the everyday and the extreme, the miraculous and the demonic. Rather, everyday magic gained its character and force from the extremes to which the stories of its origins referred. The everyday existed in a relationship with extreme events, even if these events dated from centuries before. And since categories existed as tools of governance, the possibility that an activity could be reassigned from everyday to demonic speaks more of a cultural process than of the existence of a consistent set of practices that can be referred to as everyday. The examples considered here have also shown that relations with religious authority were not solely repressive: the structures of the same church that could wreak violent condemnation provided others of its number with a source of ongoing religious vitality.

Index